Teaching Mathematics to Able Children

Valsa Koshy

David Fulton Publishers

London

David Fulton Publishers Ltd
Ormond House, 26–27 Boswell Street, London WC1N 3JZ

www.fultonpublishers.co.uk

First published in Great Britain by David Fulton Publishers 2001

British Library Cataloguing in Publication Data
A catalogue record for this book is available from the British Library

ISBN 1–85346–687–5

Typeset by Kate Williams, Abergavenny
Printed in Great Britain by The Cromwell Press Ltd, Trowbridge, Wilts.

Contents

Acknowledgements

I would like to thank all the children who have attended the special programmes for mathematically promising pupils at Brunel University, for giving me hours of pleasure. Working with them has given me valuable insights into aspects of higher ability in mathematics. I also thank them for the very special pieces of work they have produced, some of which I have included in this book.

I am also grateful to all the teachers who have attended our various in-service courses in mathematics education. I have learnt a great deal from the discussions I have had with them. This has helped me to include many practical suggestions in this book.

I am indebted to Professor Paul Ernest at Exeter University who encouraged me to write this book. He has been a constant source of inspiration for me both as the supervisor of my doctoral studies and as a mathematics educator who continues to make valuable contributions to mathematics education in this country.

I dedicate this book to my mother whose encouragement and affection means a great deal to me.

We are grateful to the following publishers and organisation for giving permission to publish extracts and figures from their publications. Their sources are acknowledged in the text:

The Association of Teachers of Mathematics (ATM)
Hodder and Stoughton Publishers
SMILE Mathematics Resources
Collins Educational Publishers

Introduction

In many years of working with mathematically very able pupils and with teachers on in-service programmes considering aspects of both the identification and fulfilment of mathematical talent, I have had some of the most exciting and rewarding challenges of my working life. In this book I wish to share my experiences with the readers.

I believe that the present time can be seen as a new dawn for the education of higher ability children in England and Wales. In the past two decades, there have been some attempts to highlight the lack of adequate provision for very able pupils. A report from Her Majesty's Inspectorate of Schools (HMI) (1978) expressed concern that in the case of the most able pupils the work was 'less well-matched' than for average and less able pupils. Not much had changed when, after 14 years, HMI voiced the same kind of concern; that many 'very able pupils in maintained primary and secondary schools are often insufficiently challenged by the work they have been set' (HMI 1992). Specific references to the lack of provision for mathematically able pupils are also included in official surveys (Office for Standards in Education (OFSTED) 1994b, 1995).

Why do I feel optimistic about a new dawn for our most able pupils? The government's White Papers *Excellence in Schools* (Department for Education and Employment (DfEE) 1997) and *Excellence in Cities* (DfEE 1999a) declare the government's commitment to education for the 'gifted'. The Schools Minister, Estelle Morris, expressed the government's objective clearly:

> The government is committed to improving educational standards for all children . . . We fail to identify many of our most able children and we don't challenge them enough.
>
> We owe to these children to help them to realise their potential. That means working with schools, parents, and local authorities to establish good

practice. We must celebrate the abilities of our most able children and
encourage them to achieve at the highest level. The attitude that gifted children
can 'cope by themselves' has let down too many young people.

(DfEE Circular 413/98)

These words are reminiscent of a statement from the USA in 1980, when the
need to do more for mathematically able pupils was recognised:

The student most neglected, in terms of realising full potential, is the gifted
student in mathematics. Outstanding mathematical talent is a precious societal
resource, sorely needed to maintain leadership in a technological world.

(National Council of Teachers of Mathematics (NCTM) 1980: 18)

The action plan for fulfilling the British government's objective, so far,
includes the appointment of school coordinators with responsibility for identifying and providing for gifted and talented children, devising school policies,
setting targets and devising strategies for the targets to be achieved in areas
designated in *Excellence in Cities* (DfEE 1999a). The government is also funding
summer schools and after-school activities for gifted and talented children in all
education authorities. Mathematically able pupils are being offered
opportunities to take 'world-class' tests and pilot projects are in place to
consider the feasibility of children taking General Certificate of Secondary
Education (GCSE) examinations in mathematics when they are 11 years old.

In this context of a new dawn there is, however, one serious cloud that raises
concern for those who are involved in teaching mathematics. There is very little
guidance available for meeting the needs of the very able mathematicians in our
schools. The questions raised by many teachers are:

- How do you identify mathematically able pupils?
- Are they all-rounders?
- What are their special needs?
- How do we know what is the right way to teach them?
- Do we teach them individually, with peer groups of similar ability, or in
  mixed ability groups?
- What kind of activities do we choose?
- How do we meet the needs of the most able within the daily mathematics
  lesson?
- What resources are available?

The purpose of this book is to provide the reader with support to deal with these
questions by offering practical strategies. These strategies are drawn from research

and from what I have learnt from the work carried out at Brunel University with teachers and pupils aged 5 to 14. The reader is invited to join me in an interactive journey exploring aspects of effective provision for mathematically able pupils. Efforts are made to provide an intelligible map for the readers; they will also be provided with opportunities for reflection on the way. By doing this, I hope the reader will feel supported and take on the challenge of providing for mathematically talented pupils with greater confidence and enthusiasm.

This book is designed to assist schools in designing a coherent policy for identifying the able mathematicians and for gathering evidence of effective provision for them. The principles and practices discussed should also provide guidance for the 'gifted and talented' coordinators in both primary and secondary schools and those who are involved in organising master classes and summer schools for able pupils.

The book is divided into seven chapters. Chapter 1 sets the context of the book and provides a general background on the topic of identification and provision for the mathematically able. With the help of case studies of pupils, a number of issues are raised. Two fundamental questions are addressed: what is mathematics and what is involved in making effective provision for our most able mathematicians? Provision for mathematically able pupils must be considered within the other developments in mathematics education such as the National Curriculum, the National Numeracy Strategy in both primary and secondary schools and the national targets set for schools. The scene is set against this background.

Chapter 2 focuses on issues of identification. Ways in which teachers can use existing structures to develop a talent portfolio for pupils are discussed. The issue of optimum conditions in which mathematical talent can be spotted are addressed and a list of characteristics, based on both teachers' experiences and international research, are discussed. Any list of attributes is only useful in so far as it can be used, and to help target provision. Such a list can play an important part in raising teachers' awareness in making practical provision.

Chapter 3 is devoted to aspects of provision for mathematically able pupils. This chapter presents a research based 'model' for provision for mathematically promising pupils with exemplars illustrating its use for achieving mathematical excellence in a sample of schools. Special features of suitable activities for use with very able mathematicians are considered, with examples of activities and pupil responses. Curriculum compacting principles and the role of higher order questioning and assessing are also discussed. In the light of the recent availability of government funding for all local education authorities (LEAs) for running summer schools for 'gifted and talented' pupils, a section on the principles and practices of a special 'master classes' programme run at Brunel University is included in this chapter.

Chapter 4 is devoted to considering provision for the mathematically able within the context of the National Numeracy Strategy. Through an analysis of the National Numeracy Framework the key messages of the strategy are highlighted, with their implications for teaching mathematically promising students. Examples of how teachers meet the needs of able pupils within the structure of the National Numeracy Strategy are provided.

Organising mathematical learning for the most able pupils in mathematics lessons is a challenge for all teachers. Chapter 5 explores different styles of organisation and considers the relative merits and drawbacks with reference to the needs of able mathematicians. This chapter also considers teaching styles in the context of provision for able pupils. The issue of differentiation is considered and some guidance is provided.

Chapter 6 looks at ways in which information and communications technology (ICT) can be used to enrich promising mathematicians. The effective use of ICT in teaching mathematics involves careful planning and evaluating learning outcomes. In the absence of these, many experiences could be viewed by pupils as superfluous, pointless and boring. The content of this chapter deals mainly with the use of calculators and computers in the context of teaching able mathematicians.

Chapter 7 looks at the effective use of resources to support high quality mathematical learning. Making effective provision for the most able mathematicians in our schools is an exciting challenge and the teacher plays a major role in meeting this challenge. Therefore, this concluding chapter of the book considers the 'desirable' qualities of successful teachers of our mathematically promising students of all ages. Aspects of designing a policy for the identification and provision for mathematically able pupils will also be briefly considered.

CHAPTER 1

Mathematically able pupils: setting the scene

The importance given to identifying and nurturing mathematical talent is not new. Several years ago, Straker (1982) convincingly stressed the importance of making provision for 'mathematically gifted pupils'. She pointed out that under the provision of the Education Act 1944, children are entitled to receive an education appropriate to their age, ability and aptitude and that children with a high level of general or specific ability are no exception. Straker maintains that:

> Gifted pupils have a great deal to contribute to the future well-being of the society, provided their talents are developed to the full during their formal education. There is a pressing need to develop the country's resources to the fullest extent, and one of our most precious resources is the ability and creativity of all children. (1982: 7)

Concerns that the needs of mathematically able pupils are not being met have been raised many times in the past two decades. HMI (1978) expressed the view that 'there was a widespread tendency to underestimate the capabilities of all groups of children, particularly the able'. The need for policies for providing for different ability levels was highlighted by HMI (1979) in the context of secondary education.

The need for making provision for children who have high mathematical ability was expressed clearly in *Mathematics Counts* (Cockcroft 1982) in the following statements:

> It is not sufficient for such children to be left to work through a text book or a set of workcards; nor should they be given repetitive practice of processes which have already been mastered.

The report goes on to say that:

> The statement that able children can take care of themselves is misleading; it may be true that mathematically such children can take care of themselves better than the less able, but this does not mean that they should be entirely responsible for their own programming; they need guidance, encouragement and the right kind of opportunities and challenges to fulfil their promise.
>
> (Cockroft 1982: para. 332)

Ten years on, concerns regarding adequate provision for able pupils continued to be raised. Alexander *et al.* (1992) pointed out that there was an 'obsessive fear' in some schools of being deemed 'elitist' and as a consequence, 'the needs of some of our most able children have quite simply not been met'. Mackintosh, Her Majesty's Inspector of Schools has this advice to offer those who feel uneasy about highlighting the needs of the able:

> There is very clear evidence that focusing sharply on what the most able children can achieve raises the expectations generally, because essentially it involves careful consideration of the organisation and management of teaching and learning.
>
> (OFSTED 1994a: 13)

More recently, schools have contacted the centre at Brunel University to seek advice when their school OFSTED inspection reports have highlighted concerns regarding the lack of challenge in the mathematics curriculum offered to children, the nature of repetitive work they are given, and over reliance on text books.

Why is it that the concerns raised in official reports have remained the same for 20 years? In a data gathering exercise during one of my in-service sessions on teaching mathematically able pupils, I asked teachers if they thought that their schools made adequate provision for their mathematically able pupils. The answer was an astounding 'no', but more interesting are the following related findings:

- Of the 74 schools from 13 LEAs represented, only three had a policy for teaching mathematically able pupils, whereas 54 schools had well defined policy statements on teaching mathematics to children with some learning difficulties. Although no one will disagree that the needs of children who experience learning difficulties must be met, it is surprising that there seem to be little appreciation that the most able also have special needs which are related to their special abilities, attributes and learning styles.

- Many of the participants felt that the publication of league tables made it inevitable that efforts and resources had to be targeted at raising the level of achievement of pupils who were borderline to reach the average Level 4 in the National Curriculum tests.
- There seemed to be an overwhelming concern about the lack of in-service support provided for teachers to enable them to explore issues of high ability in mathematics and analyse the needs of able pupils so that effective provision could be made.
- Lack of resources, especially not knowing what to do with most of what is available, was cited as a problem.
- Teachers' lack of subject knowledge in mathematics was perceived to be a barrier in making adequate provision for mathematically able pupils

This book aims to address the above concerns by offering principles and the strategies to put the principles into practice. The contents of this book should enable a coordinator to design a policy and to implement that policy with greater confidence and understanding.

Able children and mathematically able children: who are they?

Although there is no simple way of defining an 'able child' or a 'mathematically able child', it is useful to consider some issues relating to high ability in order to acquire a better understanding of how one may identify highly able mathematicians and make provision for them.

What are the complexities associated with aspects of identifying an 'able' cohort of mathematicians? First, how many children are we talking about? It is useful to consider the different perspectives that exist. In the past, it was assumed that the top 25 per cent of 11-year-olds who qualified for grammar school places were 'able'; HMI (1992) refers to 5 per cent of children as 'very able' and a model presented in OFSTED (1994b), as shown in Figure 1.1, refers to 2 per cent of pupils being 'exceptionally able'. The diversity of terminology used (George 1990) and its implications also create confusion. How do we refer to 'able' pupils? Do we refer to them as 'able', 'exceptional', 'gifted' or 'talented'? The terminology used by the government, 'gifted and talented', to identify the top 5–10 per cent of the ability range may perhaps reduce some of the tension regarding the choice of words. The term 'higher ability' pupils fits in with the 'continuum' model (Koshy and Casey 1997) shown in Figure 1.2 for referring to more able pupils; as this model facilitates more flexibility both in the identification of ability and in making provision.

Second, what do we mean by ability? Different perspectives exist.

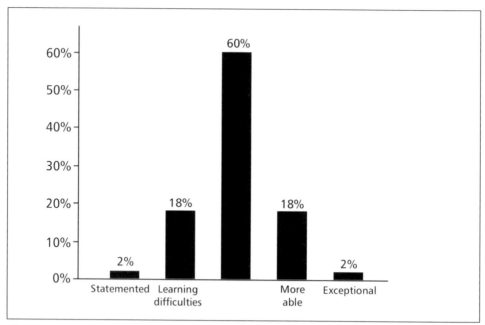

Figure 1.1 Distribution of ability in pupils in schools (OFSTED 1994b)

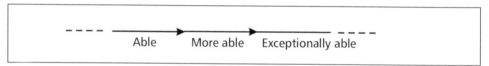

Figure 1.2 The ability continuum (Koshy and Casey 1997)

Perspectives on ability

Efforts to identify and nurture ability have been around for many decades. Early definitions of ability (Terman and Oden 1947) were based on a measure of intelligence, the Intelligence Quotient (IQ), described as the 'general intelligence factor', indicating pupils' ability to reason and make connections. In 1921, Terman and colleagues selected pupils with IQs over 140 for a study of giftedness. IQ testing is used for many purposes; for example, some schools use it for selection purposes. IQ tests are also given to pupils for diagnostic purposes. In spite of some useful purposes served by the IQ concept, many educationists and psychologists believe that a single measure of intelligence does not acknowledge the diverse talents, aptitudes and abilities of pupils. They believe that a broader concept of giftedness over a range of abilities is likely to achieve a more enhanced quality of provision for able pupils.

Ogilvie (1973) proposed that we need to consider a range of talents and abilities in children. Six categories of giftedness were listed:

- leadership
- high intelligence
- artistic talent
- creativity
- physical talent
- mechanical ingenuity.

Howard Gardner's (1983, 1993) theory of multiple intelligences describes ability in domain-specific terms. Gardner proposed seven intelligences:

- linguistic intelligence (relating to language)
- logical-mathematical intelligence (relating to mathematics and sciences)
- bodily kinaesthetic intelligence (physical – dance, sport)
- musical intelligence (music and rhythm)
- spatial intelligence (space)
- interpersonal intelligence (interpersonal skills, leadership skills)
- intrapersonal intelligence (ability to reflect on oneself).

In the context of what this book is trying to achieve, we are concerned with the second one in the list: logical-mathematical intelligence. Gardner considers two essential aspects of logical-mathematical intelligence. In a person who possesses this intelligence, the process of problem solving is often remarkably rapid. Such a person is able to cope with many variables at once and creates numerous hypotheses and evaluates them before accepting or rejecting them. Gardner (1993) maintains that logical-mathematical reasoning with the companion skill of language provides the basis of IQ tests. IQ tests assess problem-solving skills; this may explain why pupils with high IQ scores often show high mathematical ability. Aspects of high mathematical ability are discussed, in more detail, in Chapter 3.

A broader definition of 'giftedness' or 'intelligence' is recommended for the identification and fulfilment of mathematical talent by a task force in the USA. Sheffield explains:

> One of the charges to the Task Force was to determine mathematical promise in a more inclusive fashion than the traditional definition of gifted and talented that frequently only included 3 to 5% of the students.   (1999: 15)

In the task force report (Sheffield *et al.* 1995), mathematical promise is described as a function of ability, motivation, belief and experience or opportunity. None of these variables are considered to be fixed, but are areas that need to be developed so that mathematical success can be maximised for an increased number of promising pupils. The aspects listed in the report offer much food for thought for teachers in their efforts to make better and more effective provision for pupils who have the potential to do well. The report recognises that ability can be enhanced and developed and acknowledges the impact of experiences on nurturing ability. The role of motivation plays a part; careful consideration should be given to those students who may disguise their ability due to the fear of being subject to peer ridicule. Belief in one's ability as well as in the importance of mathematical success by the students, teachers and parents is also important.

Case studies of mathematically promising pupils

In the following section the reader is introduced to five case studies of pupils. I had direct contacts with four of them; the fifth was referred to me by a head teacher, and we discussed how to help the pupil on the phone. The intention of introducing these case studies is to facilitate discussion and reflection. It is likely that teachers will recognise their own pupils when they read them. After presenting each case study I have discussed it, and the reader is invited to select from the options provided in the grids that follow. This is to facilitate reflection on the complexities of identification and provision. It is likely that you will think of valid and useful options which are not included. During in-service sessions teachers work in pairs of groups to perform this task; this enables them to debate and clarify issues. You may want to record your responses on a separate sheet of paper, or use a pencil, as you may want to use these at a later stage.

Case study 1: John

In his letter to me, the head teacher wrote:

> John is 6 years old and is with a class of 7–8-year-olds at the moment. He is socially immature, but asks very interesting questions. He would like to talk to his teacher the whole time if that is possible and share his ideas with her. His interests are very different and at a very different level to that of his peers. Maybe due to this, he is dreamy and solitary most of the time. He is very clumsy and badly coordinated; there is a strange mismatch of mental and physical abilities, which creates problems for us.
>
> John is very able in most subjects, especially in mathematics. He can predict number sequences, cope with complex rules and understands time (24-hour clock included). Our mathematics adviser suggested using problem-solving

activities with him, which he finds very fascinating, but during these sessions he lives in a world of his own without any interaction. The class has 32 other children. In any case, the teacher feels unable to stretch his ability as she feels inadequate with her own mathematical ability, and then there is the question of time. We considered moving him up to a Year 5 class, but his parents and I feel that his small size and uncoordinated movements would stand in the way of him settling with older pupils.

Help! In my 30 years of teaching I have never met someone with this level of ability at his age. I want to do the best for him; but need guidance.

Which of these options would you choose for John?

Option/strategy	Yes/no	Reason for choice
Move John up to a Year 5 or Year 6 class		
Let John stay with his own class for all the lessons except for mathematics. Arrange for a mentor with the necessary mathematical knowledge and spare time to work with him three times a week		
Make arrangements to prepare John to take GCSE when he is 9 years old		
Arrange a conference with John's parents to discuss his exceptional ability and how he can be helped to do some extended work at home		
Encourage John to attend the meetings of the local mathematics club arranged by a charitable organisation		

Case study 2: Debbie

Debbie is 13 years old. She is very mature for her age and is popular with her peers in the independent school she attends. She performs well in all subjects and is at least two or three years ahead of her peers. She is among many other girls in the top set who perform well in mathematics, but her teacher, who is also the head of mathematics, feels that Debbie is a special case. She explained:

> Debbie sits with her head down in all mathematics lessons. At first I thought she was feeling bored or being rude, but later I realised she is a very polite and charming girl. Then a pattern emerged. Debbie would keep her head down obviously listening, because every time I was not sure of a step or make a mistake in my explanation she would raise her head and contribute to the discussion; most often by coming up with a more elegant or short solution or pointing out a mistake (very politely). I felt uncomfortable teaching a pupil who was working things out faster than me without the need of even seeing what was on the board. I developed such a complex that every time Debbie raised her head I thought I had made a mistake.

Which of these options would you choose for Debbie?

Option/strategy	Yes/no	Reason for choice
Move her up two years to Year 10 for all lessons		
Provide her with opportunities to explore mathematical ideas on her own including using the internet		
Register her for an Open University course in mathematics		
Give her opportunities to join national and international competitions such as Olympiad		
Let her organise a mathematics club for the junior school		
Encourage her to take the world-class tests for 13-year-olds in both mathematics and problem solving		

Case study 3: Jason

Jason is 10 years old. He has always worked very well in mathematics lessons, producing work of a good standard, which is also meticulously presented. He excels in 'number work 'and asks for homework. His teacher, Jan, received a letter from Jason's mother pointing out that her son was not being extended sufficiently in mathematics. She asked for more homework and was not very pleased when the teacher explained that Jason was doing very well and did not need to do extra homework. The visit from Jason's mother and the correspondence that followed made his teacher take a good look at Jason's mathematical performance. Jason has a good memory and assimilated all the rules and algorithms taught; he hardly ever made a mistake. He knew most of the work he was asked to do, as he had come across similar work at home. His dad taught him how to add fractions and work with decimals and ratios. Jason worked through mathematics textbooks at home and was reported to be enjoying getting the work right. He has also developed a good speed for routine operations.

After observing Jason for a week, Jan noticed that Jason liked straightforward 'sums-type' work much more than investigational work. He liked to solve problems which contained 'clue words', which suggested to him what operations to perform. Jan was impressed with his speed and accuracy and knowledge, but was not convinced he was a special case who needed special attention. Then she thought she wasn't really sure.

Which of these options would you choose for Jason?

Option/strategy	Yes/no	Reason for choice
Jason has no special talent in mathematics; his parents should be told that fact, loud and clear		
Jason's memory and his fluent work need to be acknowledged		
Observe Jason, closely, to see how he tackles open-ended investigations and problem-solving activities before dismissing him as not very able		
There is not enough evidence for assessing his special ability in mathematics. Ask for advice from the local mathematics adviser or university		
What Jason needs is to complement his numerical ability with more open-ended activities which encourage him to think why rules work. He should also be encouraged to think about other areas of mathematics – geometry and measures		

Case study 4: Richard

Richard is in Year 4; he is 9 years old. He is in the top mathematics group. You may be surprised at that because you will see very little writing in his mathematics book. He tends to do sums in his head and write just the answers down. Then he would make some attempt to set them out neatly as he knows that the teacher expects children to write down what they are doing. The answers are nearly always correct, but he uses shortcuts to work the answers out.

Richard was selected to be in the top mathematics group because of his passion for solving puzzles. These are mainly mathematical puzzles. According to the head teacher:

> he walks around with a puzzle book, records his solutions and in many cases rather than doing another puzzle, he would think of three or four different solutions to the same puzzle. Something is special about him; none of us know what it is or what to do with him.

Richard is a 'normal boy', but his passion for 'puzzle books' drew his teacher's attention to his unusual talent and ability to work with abstract ideas.

Which of these options would you choose for Richard?

Option/strategy	Yes/no	Reason for choice
Richard is a mathematical genius who can take care of himself. Leave him alone to fulfil his talent		
You need to assess his knowledge and skills to see if there are gaps in his knowledge and whether he is able to undertake calculations efficiently		
Talk to Richard's parents to find out whether they have noticed his love for puzzles and any special mathematical aptitude		
Here is boy with special talents who may have gone quite unnoticed for a long time and provision needs to be made fast		
Richard is tall and very mature; he could easily be placed with older children		
Richard is a very special case. He may need the help of a mathematics mentor		

Case study 5: Ng

I worked with 8-year-old Ng, within a group of six children who were in the top group in a mixed ability class. Ng had arrived from Vietnam only a few weeks prior to that. He could hardly speak English when he arrived, but was quickly picking up the language. The teacher commented that he seemed to enjoy mathematics lessons much more than any other subjects, which prompted her to put him with the top group. I was working with this group on a mathematical problem and made sure I explained the problem carefully and slowly so that Ng was not at a disadvantage. I was quite taken aback when he wrote down the correct solution to the problem within five minutes, while the others were still trying to make sense of what the problem was asking them to do. There were no calculations on paper, just an answer. It was difficult to get him to articulate his method, but with help he recorded some of his ideas on paper, which highlighted the use of some sophisticated strategies. Throughout that lesson and in subsequent lessons he demonstrated a lively mind and the ability to process information quickly. During one of my visits, I told him that I would be very happy to swap his 'fast' working brain for mine; he understood my comment and responded to this with a smile, and on subsequent visits asked if I still wanted to swap his brain for mine.

Which of these options would you choose for Ng?

Option/strategy	Yes/no	Reason for choice
Ng needs to get his language skills sorted out before he can be given extension work in mathematics		
Ng's language skills could be developed effectively through his mathematics work		
Encouraging Ng to work within a group will enable him to develop language skills. It could also help the other children in the group who would benefit from sharing his strategies		
Ng has mathematical ability which will remain undeveloped if it is not nurtured right away		

These case studies should help us to understand the complexity of identifying mathematical talent. They raise some questions which should help those who have the responsibility to make practical provision for pupils who are mathematically able. These questions also offer some starting points, which will be explored in greater detail in subsequent chapters. Think about the following questions:

- Is mathematically ability innate, or can it be acquired?
- Should we encourage mathematically able pupils to conform to imposed ways of working?
- What are the advantages/disadvantages of 'moving' very able mathematicians up to classes where they can work with older pupils?
- Is early entry to GCSE a desirable strategy?
- Can mathematical talent be spotted when a pupil has language problems?
- When a pupil can use correctly remembered rules and get pages of sums right, does that indicate mathematical talent or the absence of it? How can we tell?
- What practical considerations should be taken into account before establishing a 'mathematically gifted' group in the school?
- Will mathematical talent be demonstrated if an appropriately stimulating learning environment is not present?

The talented mathematician

So far, we have considered perspectives on ability and the ways in which mathematical talent can be viewed. The rest of this chapter will focus on two important questions for making effective provision:

- Why do we learn mathematics?
- What is mathematics?

The reason for much of the vague and patchy nature of provision for mathematically able pupils, I believe, is because these two fundamental questions are not always considered by those who plan provision. To set a framework for discussion of these questions, I will draw on Ernest's (2000) clear and stimulating ideas to set a framework, but readers are strongly advised to read the full version. I will also use the data I collected from over 600 (Years 4–6) pupils who attended mathematics enrichment programmes which I directed between 1995 and 2000 at Brunel University.

Why do we teach mathematics?

Before reading this section, take a moment to write down a few reasons explaining why we should teach mathematics. When I asked pupils, most of them had no problem answering this question. Most of them listed the 'usefulness' of mathematics for shopping and playing computer games and a minority listed learning mathematics to be useful for getting good jobs. Some readers may include the argument that it is a statutory requirement to teach mathematics and that it is important for children to get good results. Ernest (2000) believes that mathematics should be contributing to the education of individuals who are confident and able to see what they have learnt, sometimes in original and creative ways. Mathematics is an elegant subject and you can derive a great deal of pleasure from learning it. Bertrand Russell described mathematics as a sculpture with its austere beauty, and the National Curriculum working party had this to say:

> Mathematics is not only taught because it is useful. It should also be a source of delight and wonder, offering pupils intellectual excitement, for example, in the discovery of relationships, the pursuit of rigour and the achievement of elegant solutions. Pupils should also appreciate the creativity of mathematics.
> (Department for Education and Employment (DfEE) 1988: 3)

We can raise some important questions here. Are our pupils experiencing the joy of learning mathematics in their classrooms? Are we providing opportunities for able mathematicians to explore different lines of mathematical enquiry, put forward their theories and test them out? Are our most able mathematicians motivated enough to investigate mathematical principles for intrinsic pleasure?

What is mathematics?

Before considering the most effective ways of teaching mathematics to able pupils, we need to ask ourselves the all important question: what is mathematics and what does it consist of? Teachers' perceptions of what mathematics is will inevitably affect the way they teach the subject and, as mathematics education is going through many changes, it is even more vital that we explore this question and its implications for teaching. In order to set the framework for discussing the question 'what is mathematics?', I will draw on Ernest's (2000) paper, which also acknowledges the contribution from Bell *et al.*

(1983) and HMI (1985) in compiling the list of the following objectives for learning mathematics:

- facts
- skills
- concepts and conceptual structures
- general strategies
- attitudes
- appreciation.

Each of these objectives is now explained and the discussion focuses on how these may be useful in setting some principles for making provision for able mathematicians.

Facts

Mathematical terms, symbols and statements are fundamental to learning mathematics. Pupils need to know the correct terminology of mathematics such as names of shapes, mathematical symbols like 'greater than' (>) and 'less than' (<), the role of decimal points and how to express ratios. It is also essential that children know the units of measurement. Learning addition bonds and multiplication tables are essential for doing mathematics. Children accumulate facts or 'bits' of knowledge, which will eventually become part of their conceptual network.

Children who are very able often enjoy learning facts; they are proud of their ability to remember complex mathematical terms and usually learn these faster than other children. However, I have also met children who have high problem-solving ability being 'stuck' because they did not have the necessary factual knowledge required to solve a specific problem.

Skills

Mathematical skills include performing number operations, solving equations and learning how to use a ruler to measure lengths or a compass to draw circles. Learning a formula and using a learnt formula are also skills. What is important here is how children learn the operations or skills. With a demonstration followed by plenty of practice, children often master skills and use them competently. Skills learnt with an understanding of both the 'how' and 'why' rules and procedures work are more useful and more readily available for use in both familiar and unfamiliar contexts. This aspect is discussed, in greater detail, in Chapter 5.

Two issues are worthy of consideration in the context of teaching skills to able mathematicians. First, more able pupils take less time to learn new skills or they may already know what you are going to teach them. If they are asked to practise too much they can end up feeling frustrated and bored and consequently their motivation could be affected. Second, we need to remember that children may extract new and often more efficient procedures from what has been taught. Teachers need to encourage children to construct their explanations and procedures and acknowledge their efforts. In my experience, it is often the more able pupil who questions the effectiveness of a taught algorithm or adopts a new way of working things out. As in the case of learning facts, it is important that able pupils acquire a mastery of skills and develop fluency. Again, it is worth remembering that too much practice can kill off any enthusiasm for the subject.

Concepts and conceptual structure

A concept is an idea which becomes part of a conceptual structure. As children grow older, the different concepts they acquire make links, making their conceptual structure stronger. For example, a robust understanding of the concept of 'place value' is acquired as children understand the principles behind it. It may start with learning place names, or acquiring an appreciation of the 'grouping' concept, but children will go on making connections when they come across newer ideas, say, extension of the 'ten times bigger' idea to decimal numbers getting 'ten times smaller' to the right.

The stronger the network of concepts one possesses the more efficient one becomes in recalling and using them in other situations. Through opportunities for exploring ideas, being challenged about one's conjectures and by discussing ideas, children can be encouraged to build up a strong conceptual base. This is important for all children; but as very able pupils are often capable of processing information at a faster rate, it is particularly important to provide them with opportunities for restructuring their ideas and thoughts as part of their regular classroom work.

General strategies

The Cockcroft Report (Cockroft 1982) places 'problem solving at the heart of mathematics'. Children need to develop effective problem-solving strategies so that they can tackle both familiar and unfamiliar problems. It is quite common for children to remember clue words which help them to decide what operations to use to solve a word problem. Given a multi-steps problem or an open-ended investigation many children may not know where to start. It is useful to share some effective problem-solving strategies with children. Useful strategies for problem solving include the use of diagrams or models, breaking the problem down into smaller and more manageable parts, searching for patterns, looking for familiar parts within the problem, checking and reasoning.

As you will read in Chapter 2, children who are mathematically able are believed to possess effective problem-solving strategies. Many teachers find that this is not always the case; perhaps because the whole process is new to them. The role of teaching problem-solving strategies needs to be considered carefully. The new National Curriculum and the National Numeracy Strategy incorporate the 'using and applying' element of teaching mathematics into the whole mathematics curriculum. Will this mean the end of specific teaching of problem-solving strategies and an end to using open-ended investigations to encourage children to be creative in finding solutions, to look for patterns, hypothesise and prove ideas?

Attitudes

The way pupils perceive a subject and the experiences offered to them will influence the development of positive or negative attitudes towards that subject. All too often we hear from adults how they dislike mathematics and how they lack confidence. Fostering positive attitudes towards mathematics in *all* children is important. Quite often I have met very able pupils who have turned off mathematics because of the repetitive nature of the work and lack of challenge and interesting contexts. This needs careful thought. In subsequent chapters of this book you are offered many practical strategies which will enable you to offer an exciting mathematics curriculum to pupils. Undoubtedly, that should benefit all pupils.

Appreciation

On the first day of attendance at the Saturday programmes for young able mathematicians based at Brunel University, we ask them to comment on what they think mathematics is. Around 80 per cent of 600 children described mathematics in terms of either 'number work' or 'sums'. On the last day, when we asked them to evaluate their experiences in the programme, a good number of the pupils commented that they 'now realise' mathematics is not just about 'number work' and that they have seen mathematics in a different way. Follow-up interviews after their attendance at the classes suggest that they 'find the work more difficult, but more enjoyable'. Meeting new mathematical ideas of pattern and infinity, searching for elegant solutions and attempts at proofs appealed to most of the pupils.

The final words are, then, that although the learning of facts and practising skills are important, a mathematical diet which consists of only rule learning and memorising of ideas is deficient. Such a mathematical experience is unlikely to show our able mathematicians the aspects recommended in the non-statutory guidance:

- a fascination for the subject
- interest and motivation
- pleasure and enjoyment from mathematical activities
- appreciation of the power, purpose and relevance of mathematics
- satisfaction derived from a sense of achievement
- confidence in an ability to do mathematics at an appropriate level.

(National Curriculum Council (NCC) 1989: B11)

Take a few minutes out to think of one of your recent mathematics lessons and consider which of the above ingredients were present in that lesson, and then think of the most able mathematicians in your class and think about how they responded to your lesson.

Summary

In this chapter I have attempted to set the context of the book. Concerns expressed by educationists about the lack of effective provision for the most able have been shared with readers. Perspectives on ability were considered in order to provide a background to the issues discussed in this chapter. Using five case studies of mathematically able pupils, a number of questions were raised to facilitate discussion and reflection. Excellence in teaching mathematics to able pupils does not happen by accident; it emanates from the teachers' understanding of 'what is mathematics?' and what is involved in the effective teaching of the subject. These were addressed by considering aspects of teaching mathematics: facts, skills, conceptual structures, attitudes and appreciation of the subject.

CHAPTER 2
Identifying mathematically promising pupils

Think of the most able mathematician you have taught and write down some characteristics of that pupil. Does your list include some of the characteristics identified by Barbara (Figure 2.1), who teaches a Year 4 class? Although there is no agreed definition for a mathematically able child, most teachers don't seem to find this task difficult. Mathematically able pupils are often described by their teachers as fast learners who finish their work quickly and demonstrate a flair for doing mathematics. They are described as being able to engage in mathematical work for extended periods of time and exhibiting pleasure when solving puzzles and using construction equipment creatively. One of the issues that often arises during these discussions is the complexity in defining the term 'gifted mathematician'. Who are the gifted? Quite often I hear teachers say 'I don't think we have any really gifted mathematicians in our school, but we have some bright children'. One of the ways in which we can begin to make sense of this complex problem is to view ability as a continuum (Koshy and Casey 1997), as explained in Chapter 1 and to pay

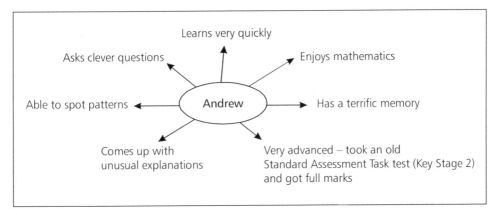

Figure 2.1 Barbara's list of characteristics of a mathematically able pupil

some attention to the nature of provision for the various groups within that continuum. For example, a child who demonstrates exceptional ability in mathematics may need to be provided with a special programme, whereas other groups of pupils who are 'able' could be grouped within the classroom or could be set just for mathematics teaching.

Identifying mathematical talent

Research literature provides some insight into what we may observe in a child who possesses special abilities in mathematics. Krutetskii, a Russian psychologist, conducted observational studies on mathematically 'gifted' pupils (1976) and offered a list of characteristics which have guided many researchers ever since. Krutetskii's list of characteristics of the mathematically 'gifted' includes:

- *Swiftness of reasoning*. The mathematically gifted are able to follow various directions of mathematical thought. They are able to offer quick solutions to problems with which they are unfamiliar by applying logic, although they may not always be able to articulate their methods to another person. It is as though they have an innate cognitive ability to process information. It is also common for these pupils to omit sequential steps which seem logical to others. Krutetskii refers to these children having 'mathematical insight'.
- The ability to *generalise* quickly is another phenomenon of the mathematically very able. While tackling a mathematical task, an able mathematician may perceive a general rule which is applicable to both the task in hand and for solving other problems. Again this ability to find a generalisation is believed to be a characteristic of very able mathematicians.
- The ability to *deal with abstract concepts* can be observed in children who have a special talent in mathematics. The most gifted may pay little attention to given data and prefer to work in an abstract form.
- The ability to *notice and make use of the mathematical structure* of a problem is another characteristic of the mathematically gifted. Mathematically gifted pupils are able to recognise and categorise mathematical problems according to their structure.
- They possess the ability to *memorise relationships* and principles of solutions from previous experience. When pupils were asked to solve a problem similar to one they had done before, Krutetskii found that pupils insisted that they had done the same problem before.

- The ability to *think flexibly* is another characteristic of the mathematically gifted. Krutetskii refers to this as economy of thought. Mathematically able pupils were able to switch from one method to another while solving problems and, as they get older, develop strategies to find the clearest, most elegant and logical solution.

Krutetskii listed two behaviour characteristics of mathematically gifted pupils which are also of importance. First, they do not tire easily when working on mathematics and are able to work for hours on mathematics without any noticeable reduction in their capability to think and compute. Second, they have the ability to see a 'host of seemingly non-mathematical phenomena through a mathematical prism'. He referred to this as possessing a 'mathematical turn of mind'.

Think, for a few moments, about pupils you have taught. Does any particular child come to mind who exhibited any or all of the characteristics listed above?

Sheffield (1994) describes a set of 'gifted and talented mathematical behaviours' that one can observe in mathematically able pupils. The following list of research based behaviours from Sheffield provides useful pointers for observation. They validate many of the attributes in Krutetskii's list. According to Sheffield's list, mathematically able pupils demonstrate:

- early and keen awareness, curiosity, and understanding about quantitative information;
- the ability to perceive, visualise and generalise patterns and relationships;
- the ability to reason analytically, deductively and inductively;
- the ability to reverse a reasoning process and to switch methods easily but not impulsively;
- the ability to work with mathematical concepts in fluent, flexible and creative ways;
- energy and persistence in solving difficult problems;
- the ability to transfer learning to novel situations;
- a tendency to formulate mathematical questions, not just answer them;
- the ability to organise and work with data in a variety of ways and to disregard irrelevant data.

Sheffield draws our attention to the fact that this list does not include the ability to compute rapidly and accurately. She maintains that while some mathematically gifted pupils may have this ability, it is not a necessary or sufficient characteristic to identify mathematical giftedness. In fact, many mathematically very able pupils are impatient with details and do not care to spend time on computation. They are often anxious to get on to the important aspects of the problems.

The two lists provided in this section, based on research carried out by Krutetskii in Russia and Sheffield in the USA, can be used to construct a single composite checklist to assess mathematical ability. Teachers often find such a list useful not only for its role in raising their awareness of the whole aspect of mathematical giftedness, but also to help them in thinking about provision. However, using such a characteristics list as the sole means of identification of mathematical talent is unwise for the following reasons.

Lack of appropriate provision

Pupils with high mathematical ability will only show their special talent if stimulating opportunities are provided. Pupils who are given a diet of repetitive exercises and closed problem-solving exercises from text books are unlikely to show their true potential. A child who is capable of detecting patterns and generalising will only do so if suitable activities are provided. Similarly, persistence and the search for elegant solutions presupposes the provision of appropriate mathematical experiences.

Fear of having to do 'extra' work

Sometimes children may want to hide their ability for all sorts of reasons. They may not want to be different or may feel that they will be given extra work which may not be stimulating. When an able pupil completes the assigned work, he or she is often given extra pages to do or harder versions of the same type of work. For example, on their completion of (often very quickly!) a page of multiplication of two-digit numbers by single-digit numbers, it is not unusual to see the pupil being given multiplications of three-digit numbers followed by multiplications of four-digit numbers. For a pupil who has mastered the process of multiplication, this strategy offers little challenge.

External factors

In the identification and fulfilment of mathematically promising pupils, we also need to consider the role of external factors. The task force which was appointed to look into the identification and fulfilment of mathematical talent in the USA (Sheffield 1999) describes mathematical talent as a function of ability, motivation, belief, and experience or opportunity. None of these variables, they maintain, are considered to be fixed, but are areas that need to be developed so that mathematical success might be maximised for an increasing number of students. This description of mathematical talent, Sheffield maintains, is based on research that acknowledges actual changes in the brain due to experiences.

Disguising mathematical ability

When identifying mathematically able pupils, it is also worth remembering that many children tend to disguise their high abilities for fear of being seen as different, losing friends and in some cases even being bullied. During a recent Saturday programme the following conversation between 9-year-old Stephen and myself transpired during a break.

Stephen: Can you keep a secret?

Valsa: Of course, I promise not to tell anyone what you tell me if you don't want me to.

Stephen: I am going to do my GCSE in mathematics next year. I have a private tutor to teach me at home.

Valsa: Are you enjoying it? I mean learning maths for GCSE?

Stephen: I haven't thought about it really.

Valsa: Why don't you want anyone to know about it?

Stephen: You see, my mum says I am very gifted. My IQ is 160. She says that makes me severely gifted.

Valsa: What will happen if others know?

Stephen: Don't you know? I will lose all my friends. Some of them won't talk to me. Some of them will ask me for answers in the class. If I give them answers – I did once – my teacher gets mad at me and if I don't then the other children will beat me up . . .

The feeling of fear, which can so clearly be seen here can hinder mathematical success. In Stephen's case, his teacher does not know how capable Stephen is so can't make suitable provision for him. She certainly does not know that he is being prepared to take GCSE at the age of 9. Stephen did pass his GCSE when he was 10; but in the classroom he worked from a text book designated for 10-year-olds. Will this experience of *twilight* secretive mathematical learning encourage him to foster the right kind of attitude to learning mathematics, which should be exciting and challenging?

Language problems

There are other factors that one needs to take into consideration when identification is being made. Children with language problems who find it difficult to communicate mathematics are not always noticed by their teachers. I have worked with children for whom English is their second language. The case study of Ng in Chapter 1, who had arrived from Vietnam six months before I met him, is a supreme example to illustrate this point. You may remember that he had neither a good command of the English language nor the confidence to make contributions to mathematical discussions in class.

Social problems

My experiences of working in schools where there is a high level of social deprivation highlights another barrier to potentially very able children being identified and reach their full potential. Quite often they lack the confidence and the basic skills needed to perform well. Lack of help at home is often pointed out as a reason for this; in many cases their parents may not have had opportunities for higher education, which makes them feel inadequate to provide a stimulating environment for discussing mathematics.

All the reasons cited above make the identification of mathematical ability a very challenging task for the teacher. Therefore, it makes sense to search for evidence of mathematical ability from as many sources as possible. In the following section I will consider some of the ways in which teachers can gather information about pupils' mathematical abilities to build up a fair and useful set of assessment evidence about their children's abilities.

Flexibility in identification

I wish to raise another point which may be valid when we consider the identification of mathematically able pupils. Identification followed by provision is not a fixed sequence.

Identification

Provision

You may, of course, notice a pupil's mathematical talent first and look for ways to make the best provision for that pupil. But we must remember that identification becomes easier if the right kind of experiences and resources are provided for the pupil. For example, you are unlikely to notice a child's ability to hypothesise and generalise if the work he or she is engaged in does not offer opportunities for demonstrating such abilities.

The next point is that any list of identified pupils should be flexible enough to allow first judgements to be revised; the 'gifted' group you initially choose is unlikely to be a 'fixed' cohort of students. Gathering information from multiple sources will not only help the identification process, it will also support the kind of flexibility we need to exercise when judgements are made.

Involving parents

As parents are likely to notice their children's special mathematical abilities, it makes sense to consult them at the time of the child starting school. Straker (1982) points out that mathematical ability often shows itself in early childhood. She gives the example of how Gauss, a great mathematician, showed mathematical promise at a young age. Gauss's teacher is said to have set a mathematical task for his class – to add up all the numbers from 1 to 99 – to keep the children busy for a long time while he went outside the classroom. Before the teacher reached the door, young Gauss had worked out the solution: $50 \times 99 = 4950$. In Chapter 3 we will look at how this solution was arrived at. A fascination for numbers, spotting number patterns or the ability to make sophisticated constructions are all indicators of mathematical ability. One parent recently told me that she was stunned when her 4-year-old son worked out the cost of three items of shopping before the sales assistant did it on the till. The child called out the total cost of the shopping to the amazement of the waiting customers and his own mother, who had previously noticed her son's unusual speed of processing numbers in his head. Providing a space on school admission forms to record special talents is a good place to start. Inviting parents to share their perceptions of their children's interests and special talents during interviews provides another opportunity. Schools need to share their commitment to nurturing talent by encouraging parents to be partners in the search for talent. Parent questionnaires and surveys are often very useful. Readers are invited to look at examples of talent and interest questionnaires for teachers, pupils and teachers in Koshy and Casey's research based model to assess 'special abilities' (2000).

Using tests

Formal assessment can often provide information about pupils' achievement. All children in maintained schools are assessed through Standard Assessment Tasks (SATs) tests. Children's knowledge and skills, as well as their competence in problem-solving processes, are assessed at the end of each Key Stage. Many schools now assess children at the end of each year. The results of these tests can provide indicators of academic success over a period of time. IQ tests are used to test potential; IQ scores are often used as a predictor of future examination success. Many LEAs now use cognitive abilities tests as indicators of potential. Standardised tests, produced by the National Foundation for Educational Research (NFER), test both basic skills and problem-solving abilities; again, these offer useful pointers for teachers to assess their pupils' abilities.

Johns Hopkins University in the USA developed a set of tests, also called SAT (Scholastic Aptitude Test), to assess pupils' mathematical and reasoning strengths. These tests are widely used in the USA and in other countries to select pupils with exceptional talent. Selected pupils are then offered opportunities for special programmes with inbuilt acceleration through the curriculum and early entry into university.

Teachers observing

Teacher observation of pupils working on appropriate mathematical tasks is one of the most effective ways of identifying mathematically talented pupils. Information regarding a child's ability can be gathered by listening to children, observing them undertaking tasks and by carefully monitoring their written outputs. Successful identification in the classroom will, however, depend on some particular, important factors. The most important factor to remember is that the right kind of opportunities have to be provided. I observed a group of ten able 5-year-old pupils working on the 'ladybird' problem (Figure 2.2) with their teacher. The children were not given any special resources, but they were told that they could select any materials they wished from a trolley which contained number lines, counters and so on.

While most children suggested that 'every other number works' or 'all even numbers will be useful', one child, Melanie, suggested that 'all numbers should work' if the ladybird is prepared to have 'half spots'. This unusual solution offered by the pupil made her teacher observe the child further in order to assess the extent of her ability.

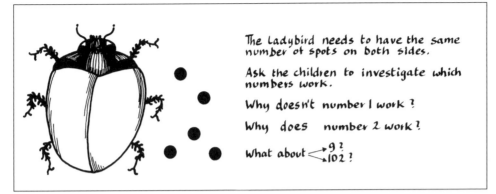

The ladybird needs to have the same number of spots on both sides.

Ask the children to investigate which numbers work.

Why doesn't number 1 work?

Why does number 2 work?

What about 9? 10?

Figure 2.2 The 'ladybird' problem (Association of Teachers in Mathematics (ATM) 1991)

The same activity was used by another teacher, who was surprised when Jason, the youngest in the group, suggested that he 'had a theory that all numbers with a zero at the end would work' because numbers 10 and 20 worked. The ability to conjecture and try a generalisation shown by young Jason provided an indicator for the teacher of his high ability. The nature of the activity you provide for children is also important when mathematical ability is to be assessed. The ladybird problem has open-ended possibilities which enabled the children to pursue the investigation at their own level.

It is a good idea to encourage children to keep their 'best' (with dates) samples of work so that the teacher can assess what progress is being made. A busy class teacher will find it easier to study portfolios of work at leisure so as to make judgements of pupils' special aptitudes and assess the reasoning behind children's output. Occasionally, it is advisable to interview a child for further information about a piece of written work so that appropriate provision can be made in future by adjusting the tasks.

Opportunities for teacher observation are also present during out of classroom activities. A mathematics club which offers puzzles and fun challenges may entice some mathematically able pupils to join in. In a non-classroom environment pupils may demonstrate their unusual talents, which can add to the teachers' wealth of information about individual children.

Using characteristics lists

A list of characteristics of mathematically able pupils, such as the ones suggested earlier in this chapter, may also be useful to raise awareness of the significance of some kinds of behaviour. We cannot expect a pupil to exhibit all the characteristics on such a list, but he or she may demonstrate an appreciable number of them. Some teachers get their colleagues to observe those pupils who show special aptitudes so as to confirm their initial judgements. Many teachers share these lists with parents, classroom assistants and others who support children in their learning.

Using peer group comments

Listening to pupils talk about their peers who have high mathematical ability is often helpful. Ask children to nominate three pupils who they think are very good at mathematics and compare their lists with your own. Experience shows that the accuracy of peer group assessment is pretty high and worthy of consideration. This way of collecting evidence is particularly useful in identifying pupils who may tend to hide their ability from the teacher for fear of being given 'extra' work!

Summary

This chapter was concerned with aspects of identification of mathematical ability. Based on international research and my own experience, characteristics of mathematically able pupils have been explored and offered as a starting point for drawing up a list of attributes which may be used by schools. Factors which may make the identification procedure difficult and the need for exercising flexibility were highlighted. In order to achieve effective identification of mathematical ability, we need to draw on a range of evidence; parents' views, test results, teacher observation notes and peer group comments all help to build up a picture of a pupil's mathematical talent.

Two points are worthy of reflection at this stage. First, using the criteria discussed in this chapter, make a list of children you know or teach who you think are able mathematicians. Second, make an appraisal of what provision you currently make in order to meet their needs.

CHAPTER 3

Effective provision for mathematically able pupils

In Chapter 2 we explored ways of identifying pupils who possess special abilities in mathematics. In this chapter we will consider how to make the best possible provision for them. At the outset, we must remind ourselves that pupils who are identified as having high mathematical talent may vary a great deal in the degree of their ability and attitudes. A promising mathematician who enjoys the subject and has developed a passion for, say, problem solving can be given challenging work; this kind of pupil may not require much teacher intervention. On the other hand, a pupil who demonstrates high potential but is lacking in motivation may need more teacher input as well as a greater amount of encouragement and coercion before she or he gets fired up with enthusiasm. Such pupils may also need a structured programme to get them motivated initially. In spite of the inevitable differences shown by the pupils we can set some principles which can help teachers when considering provision. In setting these principles, I have taken into account both the attributes of able mathematicians and the views of practising teachers who trialled some activities for a research project on curriculum provision for able pupils. The principles discussed briefly here will be revisited throughout the book.

Principles of provision

Mathematically able pupils need to have their talents identified and acknowledged

In this respect these children are no different to other children and adults who respond more positively when their talents are spotted and acknowledged. In the case of very able pupils this is particularly important. Teachers may avoid asking the brightest child who puts their hand up all the time to contribute to the discussion, in order to give a chance to the other children. This can naturally be quite upsetting for an able pupil, whatever the age. Thirteen-year-old Hamash once told me that he no longer put his hand up because 'I never get asked – I might as well give up'. In the case of an able pupil who has a lower level of motivation this is particularly important.

Mathematically promising pupils should be exposed to advanced concepts which they are often capable of exploring

Mathematically promising pupils are almost always fast learners. They process information faster and are capable of learning complex ideas earlier than their peers. Quite often they already know what the teacher is planning to teach the class. Repetitive work and too much drill and practice should, therefore, be kept to a minimum for these children. The more able a pupil is, the less appealing this type of activity becomes. Special care should be exercised when using textbooks with these children. Studies conducted in the USA (Sheffield 1999) highlighted the following, which is also significant in the UK:

> Mathematics text books in the United States tend to cover large numbers of topics at relatively shallow level and repeat the same topics for years. . . . This is especially detrimental to good mathematics students who have already mastered the content of the mathematics programme and are bored with the repetition.

Mathematically able pupils benefit from divergent tasks

Closed tasks which require single solutions are not likely to encourage thinking in depth. Although all children enjoy getting a good number of ticks in their books, mathematical teaching which consists of only closed tasks may fail to challenge the able child. Most of the very able young mathematicians I have worked with showed persistence and stamina and often they resented time constraints and imposed restrictions, especially when enjoying a topic or pursuing a line of enquiry. During the mathematics master classes at Brunel University we have come across children who would work for two hours on an investigation and still show disappointment when they are asked to stop. Mathematically able pupils are also more capable of reflecting on their own leaning; this obviously needs time.

Able mathematicians benefit from communicating with peer groups of similar abilities

My experience of working with groups of able pupils has shown that they benefit from working together, in terms of both learning mathematics and socialising. They learn from each other and challenge each other. They can share their interests and achievements in the subject. Most of the able pupils I know show curiosity and ask more questions. In a busy classroom the teacher may not always have the time to respond to all the questions, so working in peer groups of similar ability gives children an opportunity to enhance their understanding of mathematical ideas.

Mathematically able pupils are often more capable of analysing, reasoning and evaluating

The implication of these capabilities is that the mathematics programme should provide them with opportunities to analyse ideas and extract principles, look for patterns and relationships, reason and consider alternative solutions. Most textbooks and activity sheets are designed with average pupils in mind; they don't always focus on these processes. Here is a task for you to complete, individually or with colleagues. Select a few pages from three or four textbooks and mark any ideas which require children to analyse information or exercise reasoning.

A key concepts model for mathematical provision for able pupils

When considering the topic of teaching able pupils, most people make references to identification and provision. In the search for a viable and effective strategy for effective provision for able pupils, the Able Children's Education Centre at Brunel University has focused on the construction of a theoretical model to provide practical support for teachers. Casey's (1999) key concepts model for mathematics has been successfully used as a basis for 'good practice', for providing for able mathematicians and setting up centres for excellence in mathematics in LEAs. In one LEA this model provided the basis and for setting up a centre for excellence for mathematics in a primary school. Why have a subject-specific model? Gardner's view of domain specific intelligences (1983, 1993) and our own experiences and beliefs have led us to consider subject-specific provision as a practical way forward to both identifying and nurturing ability.

Content-specific models of provision are gaining much support among leading experts in curriculum provision for able pupils. In a clear and convincing voice, Van Tassel-Baska (1992), a leading expert in the USA on curriculum provision for gifted pupils, lists the reasons for moving to a content based instructional model for the gifted. Van Tassel-Baska's arguments are both logical and practical:

> Schools are organised by content areas, and to deviate significantly from these areas is to be outside a predominant organisational pattern that aids communication on gifted issues within the school system. It also provides the natural context for planning the curriculum, because of school systems, even those with self-contained programs for the gifted, are obliged to show mastery of basic skills for gifted in these subject matter areas . . . the impact of programmes for the gifted is limited by ignoring content.   (1992: 2)

Van Tassel-Baska argues that knowledge at a social level too is organised in discipline-specific ways. She puts forward a convincing argument for her view by citing the example of Nobel prizes in specific disciplines.

In our ongoing dialogue with teachers, they have often articulated the need for subject-specific provision for able pupils as opposed to being guided by 'generalist' principles for provision. In Britain, where we have a National Curriculum in specific subjects and a National Numeracy Strategy, consideration of subject-specific provision for the mathematically able does seem a natural and sensible way to proceed.

The key concepts model for learning mathematics has been devised in order to support teachers with a framework for provision. It emphasises the need for a balance of learning facts and skills, at the same time offering some freedom of approach. Learning programmes which follow the model often lead to the discovery of previously unsuspected mathematical talent.

The basic features of the model consist of three sets of interacting components (Figure 3.1). Casey (1999) explains them as:

- a set of five components relating to the learner's outcomes and dispositions – facts, skills, fluency, curiosity and creativity;
- a set of five components relating to the methodology of the specific subject (in this case the subject is mathematics) – *algorithm, conjecture, generalisation, isomorphism* and *proof.*

Other facets of the learner's situation, considered from a somewhat sociological perspective, are also being incorporated into this model, but are not described here. Papers will provide further illumination of these aspects in the near future.

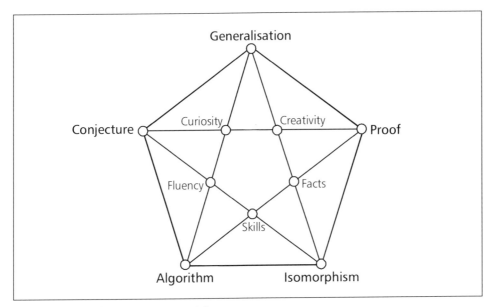

Figure 3.1 The key concepts model

The first set of five components is represented by the five vertices of the inner pentagon. They are selected to ensure a balance between the discipline and the practice needed by the pupils to acquire facts and skills and develop fluency and at the same time give them the freedom to pursue ideas which arouse their curiosity, which should help to develop a capacity for creativity.

Why are these five facets included? Pupils need to learn *facts* and *skills*, as discussed in Chapter 2. They need to recall facts quickly and develop speed and confidence. They need to learn mathematical vocabulary and become competent in using symbols and signs. The seriousness of pupils' lack of numerical *fluency* was highlighted by the London Mathematical Society (1995). One of the difficulties experienced by students following mathematics courses at universities was 'with fluency and a serious lack of essential technical facility – the ability to undertake numerical and algebraic calculation accurately' (London Mathematical Society 1995: 2).

One of the reasons for setting up the National Numeracy Strategy (1998a) was to develop pupils' fluency in numerical work. Without knowing facts and being able to exercise skills, it would be impossible to be engaged in mathematical enquiry.

Without the inclusion of *curiosity* and *creativity*, Casey maintains that the curriculum becomes a cage. Quite often for a capable mathematician the curriculum, with an over-reliance on textbooks and ready-made worksheets, can become so. Casey's memorable words in support for these children were 'Caged birds do sing, but what would be their song if they were sometimes allowed to fly?' (1999: 14). A creative mathematics curriculum will encourage questioning and let pupils pursue different lines of enquiry.

Now consider the outer set. *Algorithms* for performing basic operations are essential for learning and doing mathematics. Again, learners should be allowed the freedom to use their own algorithms, the ones they feel most at home with. The National Numeracy Strategy endorses this idea. Although it is necessary to teach standard algorithms, too much practice of these algorithms often leads to boredom. A learner who has acquired competence in carrying out number operations should be encouraged to use and develop personal algorithms to pursue interesting mathematical investigations. A study of adding 'consecutive numbers' can provide a very meaningful context for focusing on conjectures. The following *conjectures* were made and investigated by 9-year-old Daniel:

- 'When you add two consecutive numbers, the answer is always odd.'
- 'When you add three consecutive numbers the answer is always odd.'

Daniel revised his conjecture after trying different numbers to 'when you add three consecutive numbers the answer is odd only if you start with an even number'. Following this investigation, Daniel wrote a booklet entitled 'What everyone should know about consecutive numbers'.

A pupil who conjectures that 'the sum of two odd numbers is even' can be encouraged to *generalise* and prove the statement. Even young learners are capable of producing some kind of proof, perhaps in words. One such proof can be the following:

Let the two numbers be 8 and 9

$8 + 9 = 17$

And now let the two numbers be 23 and 24

$23 + 24 = 47$

So if the first of the two consecutive numbers is odd or even, the sum of the two numbers is odd. This could be taken to be the conclusion. Evidence for it being true has been provided by offering two particular examples: starting with an even number and starting with an odd number. This makes the conclusion plausible, but mathematically it can be proved algebraically as follows:

Let the first number be n.

Then the next number will be $(n + 1)$.

The sum of the two consecutive numbers is $n + (n + 1) = 2n + 1$, which must always be odd ($2n$ is a double, which must be even, so $2n + 1$, the number after $2n$, must be odd).

The concept of *isomorphism* gives mathematics its power and it gives the learner some excitement. Observing a child realising that seemingly different situations have the same underlying mathematical structure is a worthwhile experience. Krutetskii's research, which was described in Chapter 2, in fact identified that mathematically able pupils do have a talent for appreciating the underlying structures of the subject. A pupil who points out similarities of the present experience with situations encountered in the past gives us an indicator of his or her high potential.

Designing and selecting activities for able mathematicians

The key concepts model described in the previous section can be used as a basis for designing and selecting tasks for able learners. At this point, I would like to emphasise that the features of activities listed below constitute 'good practice' for all children and can be designed for use with the whole class. What are the features we should be looking for? I have listed some questions which one could ask in selecting activities, which have been found useful by practising teachers.

Does the activity offer interest and challenge?

Challenge is an interesting word. When I ask people what they think challenge is, I often get a definition involving the notion of 'something difficult'. The statement that 'able pupils need challenges' is too much of a simplification. It is true that they need to be provided with tasks which stimulate their thinking, but this does not mean that you have to search for difficult tasks. Investigating the properties of numbers can start as a simple task which can be extended to more sophisticated levels. It is all important that the tasks should be presented in an interesting way. I have sometimes found younger, bright mathematicians being demotivated when they were given worksheets or photocopied pages selected from textbooks designed for older pupils. It is true that these worksheets expose them to difficult and more challenging concepts. But if pupils are put off by the presentation they are unlikely to respond with their best efforts.

Contexts are important too; realistic contexts appeal to children. However, I have found that mathematically able children get very excited both about tasks which are based on pure mathematics and about tasks presented in realistic contexts. Quite often the way teachers present the tasks influences pupil response. Two examples should illustrate the point I am trying to make.

Furnish a bedroom

During an enrichment programme, I used a task entitled 'Furnish a bedroom'. It was a problem which was originally designed for a curriculum research project with groups of children (Casey and Koshy 1995). The children were given a list of items to choose from on a page of a catalogue which contained items of furniture, soft toys, electrical goods and other luxury items. Some prices were reduced by 5 per cent, some by 12.5 per cent and there were discounts on other items. Pupils were told that they should furnish and create the best possible bedroom with a budget of around £600. They had to show how they would spend the money. They also had to draw a plan of the bedroom, to scale, to show where the furniture would go. No calculators were allowed!

As well as being introduced to the idea of optimisation, children had to use ideas of scale, percentages, discounts and estimation. Decision making and communicating were also involved. The idea motivated the children so much that they stayed on during their breaks to carry on with the task. The children were 7- and 8-year-olds, many of whom had not previously met the ideas of percentage discounts, scale drawing and so on. The motivation was so high that they were very keen to learn about how to work out percentages. Playground talk focused on the 'bedroom problem' and parents pointed out that children were discussing changes to their own bedrooms at home and working out percentage reductions of items in newspapers and shop windows. The challenge was there, but an interesting context and the presentation of the problem made all the difference. Teachers who trialled this idea commented on both the positive attitudes of pupils and on how they were motivated to learn new ideas and how to apply them.

Smarties

Similarly, the 'Smarties' project, which started as a whole-class activity with 7-year-olds, kept all the children interested for weeks. The more the able pupils were given the responsibility for investigating, the more complex were the questions they pursued. The class teacher brought in a selection of tubes of Smarties and used them for her carpet discussion. First, children were encouraged to ask questions about the tubes of Smarties. Initially children were reluctant to ask questions and one commented 'it is the teacher who asks questions, not the children'. This myth had to be dispelled in order to make the lesson worthwhile and challenging. Questions soon flowed afterwards:

- 'How many Smarties are there in a tube?'
- 'Do all the tubes have the same number of Smarties?'
- 'How many colours are there in a tube?'
- 'Do all tubes have the same number of each colour?'
- 'Is there a relationship between the colour of the lid and the number of Smarties of a particular colour inside the tube?'
- 'Does the letter inside of the lid say anything special about the particular tube?'
- 'How much would ten tubes of Smarties cost?'

Groups of children were asked to investigate different questions; the higher ability pupils were given the more complex aspects to investigate and were encouraged to pose more questions of their own, which included the cost and weight of one Smartie. Calculators sometimes had to be used to find the solutions to the problems. The more able ones were also given the responsibility of putting up a display about Smarties. They used ICT packages for undertaking some of the investigations and gave a presentation, 'All about Smarties', in assembly.

In both examples, children learnt and/or practised mathematics meaningfully and with much excitement. According to their teachers their positive response to the challenges were mainly due to the motivating contexts and the teachers' own enthusiastic introductions. I have no doubt that their teachers' enthusiasm also contributed to the successful outcomes.

Does the activity have potential for different levels of outcomes?

The ultimate aim of all good programmes in education must be to encourage all pupils to reach their full potential. One way of achieving this is to make sure that you include activities which have different levels of possible outcomes. This will maximise the challenge for every pupil and then those pupils who are more able will extend the task further by teacher suggestions or by their own interest. It is worth bearing in mind that it is not necessary that all children should start at the same level. I will illustrate this with an activity.

Handshakes

'If one morning all the children in this class shook hands with each other, how many handshakes would take place?' The handshake problem can be given to the whole class. The teacher may decide to ask one or two groups of children to find out how many handshakes would take place at their table with four children. This easier version of the problem will enable all children to participate and achieve successful outcomes. Various levels of outcome are possible. The children could:

- find the solution for the number of handshakes for people by drawing pictures or by actually shaking hands and counting the number of handshakes which took place (answer: six handshakes);
- work systematically and use a table to record results to find out how many handshakes would take place.

In the second case, the table they would come up with would be:

Number of children	Added handshakes	Total handshakes
1	0	0
2	1	1
3	1 + 2	3
4	1+ 2 + 3	6
5	1 + 2 + 3 + 4	10
6	1 + 2+ 3 + 4 +5	15

and so on. By spotting the way the sequence works, it can be worked out that for seven people it would be 21, for eight people it will be 28, for nine people it will be 36 and for ten people the total number of handshakes will be 45.

For a larger number of people, working out the total number of handshakes using this method would be laborious, even with the aid of a calculator. The teacher can suggest, to those who are capable of taking it further, that they look for an easier way to work out the number of handshakes. Is there a connection between the number of people and the number of handshakes? Can they spot that relationship? Your more able pupils may take up the challenge and establish a rule and explain how they arrived at a generalisation.

If there are n children the number of handshakes will be given by:

$$\text{number of handshakes} = \frac{n(n-1)}{2}$$

You could ask the children to explain the rule to the rest of the group or class.

Does the activity provide opportunities for generalisation and formulating proofs?

The handshakes activity provides pupils with opportunities to look for patterns and explore the possibility of generalisation. I observed a class of 8-year-olds tackling the 'towers' problem.

Towers
This tower, as you can see, is two cubes tall.

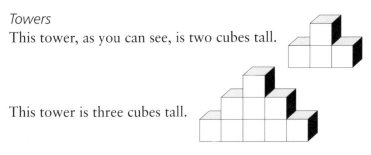

This tower is three cubes tall.

- Think about the tower you could build if you had 20 cubes. What height would that tower be?
- What if you had 150 cubes?
- What if you had 1000 cubes?

It was interesting to note the different methods and strategies used by the children.

The teacher, Allison, does not 'believe in' providing children with a predetermined set of resources when they are working out problems. The resources are always made available for them to use if and when they feel they need them. She explained her reasons:

> I used to give all kinds of resources to children and they always felt they should use them because the resources were on the table. Then I came to realise that some children may not need the resources at all and even if they do, it would be good to let them make the decision of what to use and how to use it. In this particular case I could have given them bricks at the start, but I didn't. One of the ways you can differentiate is by the availability of resources. Only two out of the four groups went and collected cubes to investigate the tower problem.

A study of how children tackled this problem is analysed below:

- One group used cubes to build two more towers and found out that they could build a '4-tall tower' with 20 cubes and have four cubes left over. The 'left over' concept did puzzle them a bit, but they accepted it. Allison asked them to build a few more towers and record the results, which they did successfully.

- Another group also built two more towers and started writing down the results. They used a table as 'they are trained to do'. From the table, they could work out the solution to the first question: they could build a '4-tall' tower using 20 cubes with four cubes left over.

Height of the tower	Number of cubes used
2	4
3	9
4	16
5	25

- The same method of recording was used by another group, who also spotted a relationship between the height of the tower and the number of cubes used and managed to produce a solution for all three parts of the question. You 'times the height of the tower by itself to work out the number of cubes'. Allison introduced them to the idea of 'squaring' a number and also introduced them to the square root key on the calculator. The lesson became more interesting after this. A generalisation was suggested which they had to test and prove and then share with the rest of the class.

Mystic rose

The 'mystic rose' problem (Figure 3.2) similarly offered opportunities to a group of 10-year-olds to investigate and find a rule.

Points	Lines	
2	1	$1 + 0$
3	3	$2 + 1 + 0$
4	6	$3 + 2 + 1 + 0$
5	10	$4 + 3 + 2 + 1 + 0$
$\vdots$	$\vdots$	
n	$\dfrac{n(n-1)}{2}$	$(n-1) + (n-2) + \ldots + 1 + 0$

The children were delighted to find a link with the previously investigated 'handshakes' problem.

Isomorphism had been learned by experience rather than by abstract definition.

Join points on a circle to other points.

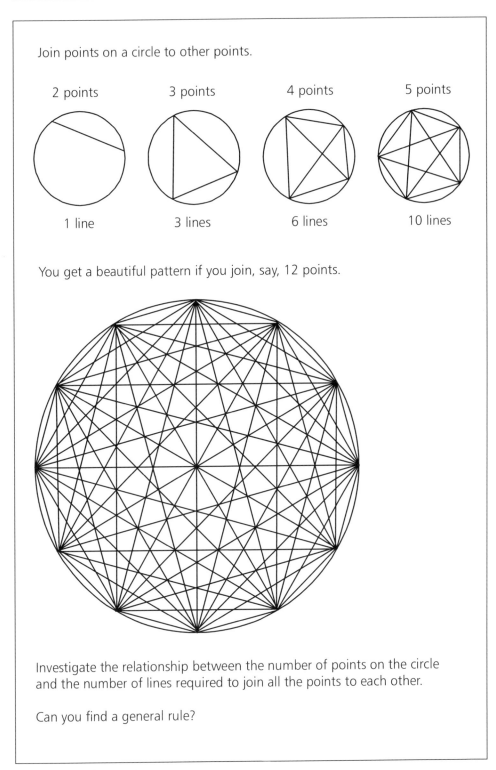

Investigate the relationship between the number of points on the circle and the number of lines required to join all the points to each other.

Can you find a general rule?

Figure 3.2   The mystic rose

Does the activity encourage reflection and purposeful recording?

The work presented here (Figure 3.3) is from 9-year-old Emma, who investigated the relationship between the number of sides and the number of diagonals of polygons. This problem has similarities with the 'mystic rose' problem. Emma's teacher, Jim, encourages children to keep a record of their thinking and progress. Work started in a lesson can be continued at other times or be taken home. Jim believes it is important for children to have an extended period of time to think and reflect on their ideas and Emma's work certainly endorses that.

Does the work offer opportunities for children to show what they are capable of?

Finally, it is important to set high expectations for all children and it is also very important to set work which provides able mathematicians opportunities to show their true potential. As I have mentioned previously, it is useful to remind ourselves that most textbooks and activity sheets are designed with the average pupil in mind and that most able pupils should be able to complete these faster than their peers. Observing pupils doing routine mathematics will help the teacher to assess their competence in many aspects of mathematics, but in order to identify pupils with mathematical promise they have to be exposed to challenging work. These challenges may be given to groups within the class, to children who are in the top-set or to individual children. Many of the examples given in the previous sections are suitable tasks for observation.

Curriculum compacting

During my visits to schools, I have observed children who look bored and disinterested in the mathematics lessons they are attending. This is a problem which is also increasingly being highlighted in OFSTED inspection reports of schools. OFSTED reports raise the issue that in many classrooms very able pupils are being asked to carry out mathematical work which is either too easy or repetitive. This may be for several reasons. Preparation for tests to maximise the number of children who reach 'average' expectations may be one and undue reliance on textbooks which have very new material at different stages of schooling may be another. Whatever the reason, pupils who are not sufficiently challenged can lose their motivation and achieve less. *Curriculum Compacting* (Renzulli 1994) is a system recommended by the authors to adapt the regular curriculum to meet the needs of the gifted students by either

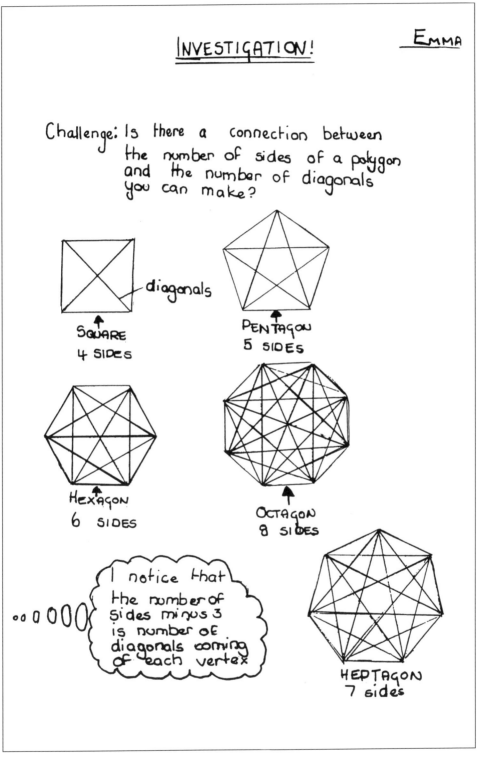

Figure 3.3   Emma's investigation (continued overleaf)

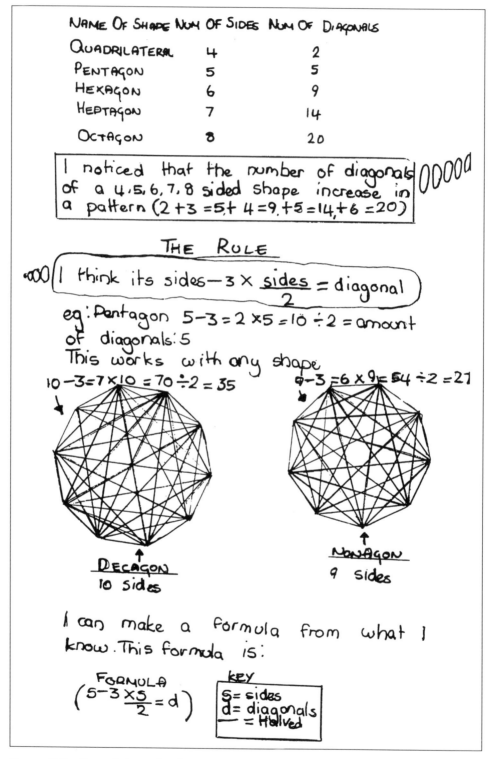

NAME OF SHAPE	NUM OF SIDES	NUM OF DIAGONALS
QUADRILATERAL	4	2
PENTAGON	5	5
HEXAGON	6	9
HEPTAGON	7	14
OCTAGON	8	20

I noticed that the number of diagonals of a 4, 5, 6, 7, 8 sided shape increase in a pattern (2 + 3 = 5, + 4 = 9, + 5 = 14, + 6 = 20)

THE RULE

I think its sides − 3 × $\dfrac{\text{sides}}{2}$ = diagonal

eg: Pentagon 5 − 3 = 2 × 5 = 10 ÷ 2 = amount of diagonals: 5

This works with any shape

10 − 3 = 7 × 10 = 70 ÷ 2 = 35   9 − 3 = 6 × 9 = 54 ÷ 2 = 27

DECAGON
10 sides

NONAGON
9 sides

I can make a formula from what I know. This formula is:

FORMULA

$\left(5 - 3 \times \dfrac{S}{2} = d \right)$

KEY
S = sides
d = diagonals
— = Halved

Figure 3.3   Emma's investigation

eliminating work which has already been mastered or to streamline work that can be learnt easily by the pupil. The time released by this process can be used for enrichment activities which provide opportunities for thinking in more depth or learning more complex and exciting ideas. Curriculum compacting can be done during planning time. Some teachers actually involve able children in the process by asking them to explain why and how parts of the textbooks, or the content learnt by most children, may be cut out.

Giving children opportunities for higher levels of thinking

Able pupils, like all other pupils, need to have opportunities to be engaged in higher levels of thinking. Gifted pupils often demonstrate superior cognitive capabilities. Therefore, it is our duty to encourage them to be effective thinkers and creative problem solvers. In this section I will introduce a useful framework which is used in many schools, both in the UK and elsewhere, for planning activities in all subjects. I have not seen many examples of using Bloom's taxonomy of educational objectives to plan mathematics, so I have taken on that challenge.

In his *Taxonomy of Educational Objectives: the Cognitive Domain* (1956), Bloom established a classification of levels of cognitive functioning, which has been widely used for conceptualising levels of thinking. A summary of the six levels – knowledge, comprehension, application, analysis, synthesis and evaluation – is provided here, with examples from the mathematics curriculum.

Knowledge

This level is concerned with learning facts. Number names, symbols and names of shapes are examples. This is the basic level where a learner is asked to list and remember facts. Able pupils may already know many of the facts. But it is worth remembering that all children need to learn them. A firm knowledge base is necessary for mathematical explorations. Sometimes facts can be learnt without a conceptual understanding of what is involved when the acquisition of knowledge has been through a transmission model of learning – the teacher 'telling' the pupils or a book providing information. This model of learning is unlikely to lead to knowledge which can be applied when the need arises. In a mathematics lesson at this level, children:

- will be able to name and identify triangles, squares and circles;
- will know the units of measurement (centimetres, kilometres);

- can recollect their multiplication table bonds;
- can recall and record the symbols for the four operations, equals, percentages and so on.

Comprehension

This level is concerned with showing what knowledge has been acquired and being able to explain or retell what has been taught. This means either summarising what has been told or answering questions based on what has been taught. At this level a higher level of understanding is achieved through explanations and instruction. In the classroom, teachers ask direct questions for pupils to respond. In a mathematics lesson at this level, pupils may be able to:

- describe a shape and its properties;
- perform an algorithm;
- describe the composition of a two-digit number as tens and units;
- work out a simple problem which involves the use of a known algorithm;
- complete a number pattern;
- use a ruler to measure a straight line;
- collect simple data.

Application

At this level, pupils can be asked to use and apply what they have learnt in both practical and problem-solving situations. This level of thinking requires more advanced cognitive processing. The application level involves more explanations and making more conceptual links. In a mathematics lesson at this level, pupils should be able to:

- make up number stories;
- use methods to solve simultaneous equations;
- apply rules to situations;
- select methods to solve word problems;
- organise data and produce bar charts.

Analysis

Activities at this higher level involve looking at situations and breaking them down into components and perceiving relationships among the parts. Analysis also involves detecting patterns and structures and underlying principles. This level involves a more complex level of thinking leading to personal discoveries and generalisations. It is important for teachers to provide extra guidance for pupils, at this stage, when they are forming and internalising ideas. In a mathematics classroom at this level, pupils could be expected to:

- identify and continue patterns;
- explore and establish the underlying structure of mathematical ideas (e.g. place value);
- understand the interrelationships between concepts such as fractions, decimals and percentage;
- begin to hypothesise, conjecture and try to generalise;
- try to establish proofs;
- analyse textbooks;
- write a text page on a mathematical topic.

Synthesis

This level of thinking involves more creative thinking. It involves looking at something in a different way, posing new questions, hypothesising and creating new insights. 'What if' type questioning is a feature of this stage. This level encourages thinking styles which produce more original and creative ideas. Pupils may design something new or refine an existing idea. Planning activities at this level is challenging for the teacher too, as new ideas are not easily transported from textbooks and worksheets. Personal theorising can take place and unique products may be created. At the synthesis level pupils will still need to use the facts and skills they have acquired at the previous levels in order to produce superior and worthwhile products. In a mathematics lesson at this level, children should be able to:

- create a new number system;
- ask 'what if' to create new situations;
- generate new and unique solutions;
- write a few pages of their own version of a mathematics textbook;
- challenge existing ideas and methods and suggest more effective solutions;
- work in different bases.

Evaluation

This is classified as the 'highest level' and it is not surprising that this is so. Personal reflection, self-assessment and metacognitive processes are present at this stage. It involves a process of making judgements based on a sound set of principles and criteria. Again, the knowledge base and the skills acquired at the other levels are needed to make a sound appraisal of something. Presentation of products to others, reviewing the effectiveness of their own and other people's efforts are also features of activities at this level. At this level, mathematical activities can include:

- making an appraisal of solutions and judging their relative effectiveness;
- evaluating the textbook treatment of topics (say, probability);
- a self-assessment of one's own learning;
- recording steps in one's thinking;
- examining a proof formulated by oneself or someone else;
- judging the superiority of certain algorithms or procedures over others.

Bloom's taxonomy provides us with a framework for asking more challenging questions and planning more stimulating activities for the mathematically able. As able pupils are potentially able to think at higher levels, this framework offers a useful framework for planning purposes. I would like to reiterate that able children need to have the same instruction received by all children for learning facts and skills, but it is likely that they will master these very quickly. The teachers' task is to plan, in advance, activities which encourage them to analyse situations critically, synthesise new ideas and evaluate the effectiveness and quality of their thinking processes as well as their products.

Higher order thinking skills

An effective mathematics curriculum will encourage pupils to be involved in higher levels of thinking. As well as learning facts and procedures it will encourage skills of communicating , reasoning, decision making, hypothesising, refining ideas and creating new and original thoughts. Resnick (1987) listed the following higher order thinking skills, which are desirable in a mathematics curriculum and are worth bearing in mind. Higher order thinking:

- is *non-algorithmic* – the path of action is not fully specified in advance;
- tends to be *complex* – the total path is not 'visible' from any single vantage point (mentally speaking);
- is often what yields *multiple solutions*, each with costs and benefits, rather than unique solutions;

- is the product of *nuanced judgement* and interpretations, not pre-determined;
- is the application of *multiple criteria*, which sometimes conflict with one another, not a set of simplified criteria;
- often involves *uncertainty* – not everything that influences the task at hand is known;
- is concerned with the *self-regulation* of the thinking process – we do not recognise higher order thinking in an individual when someone else calls the shots at every step;
- is indicated by *imposing meaning*, finding structure in apparent disorder;
- is *effortful* – there is considerable mental work involved in the kind of elaboration and judgement required.

It would be a useful exercise to consider the nature of your mathematics lessons and judge which of these higher order thinking skills is being displayed in practice. In the light of Resnick's list we can raise some valid questions. Is most of the mathematics work in classrooms largely algorithmic? Are our examples usually standard, with visible directions given? Do the problems we set for children allow for multiple solutions or are they designed to expect a single solution? How often do we expect children to make judgements or produce interpretations? Do we force children to apply a single criterion? Do we always provide all the information needed? Do we allow children some freedom of direction or do we impose our thinking on them? Do we encourage children to think mathematically and create order in a set of unconnected and chaotic situations? Does our mathematics work consist of problems with single solutions which require little effort?

Creating a problem-solving culture in mathematics

In the previous section we discussed the role of thinking skills in planning the mathematics curriculum. One of the practical ways in which thinking skills are utilised is through problem solving. Thinking is a mental process which helps a person to use and apply his or her knowledge to new situations. It involves reasoning, hypothesising and decision making. These processes are vital for problem solving. As discussed in Chapter 2, mathematically able pupils possess a greater capacity to engage in these processes. Straker quotes an infant teacher:

> the child with a quick mental facility, who can see pattern and build on known facts to help problems, is showing indications of mathematical

potential. This kind of child copes well with formal practice work, as do most children, but it is when they are given problems to solve that one sees a difference.   (1982: 1)

In my own experience, children who are very able in mathematics demonstrate a flair for problem solving. Some have become addicted to solving puzzles and activities which involve logic and reasoning.

Problem solving has always been seen as an important part of learning mathematics. The Cockcroft Report (Cockcroft 1982) describes problem solving as being 'at the heart of mathematics'. The 'using and applying' component of the National Curriculum (DES 1991) encourages the strategies which support problem solving. Three areas were listed. The first is about *making and monitoring decisions* to solve problems with regard to the selection of materials and procedures to solve problems. The second area is concerned with *developing mathematical language and communication*, which involves oral communication and recording and presenting problem-solving activities. The third is *developing skills of mathematical reasoning*, which is concerned with reasoning, checking and justifying results.

The National Numeracy Strategy has included solving problems as one of the strands to be taught to children. The strategy also places much emphasis on oral communication, reasoning and checking. More recently, the government has announced the introduction of world-class tests in mathematics and problem solving (DfEE 1999a) as part of the provision for gifted and talented children.

One of the ways in which we can provide for gifted or able mathematicians is by providing them with opportunities for problem solving. In this context, we need to ask what we mean by problem solving. George Polya, whose ideas about problem solving guided most mathematics programmes from 1945, gives us a definition:

> To have a problem means to search consciously for some action appropriate to attain some clearly conceived but not immediately attainable aim. To solve a problem means finding some sort of action. In mathematics the ability to solve problems is not just knowing some straightforward rules which makes one a better problem solver.

To be an effective problem solver in mathematics one needs to possess a degree of independence of thought, a capacity to make and monitor decisions and produce creative solutions. Problem solving which involves the above features should naturally provide challenges for our able mathematicians.

Can we train children to be effective problem solvers? I can partly answer this question, based on experience. I certainly know that you can train children to think more efficiently, be persistent, be systematic and to accept mistakes as a positive feature of their learning. Several years ago, when I was an advisory teacher for an LEA, I initiated problem-solving groups in six schools for bright children. Every week they had a session with me when we would solve problems. Initially, most children did not know where to start. They would give up very quickly if they did not know exactly where they were going. In time, however, they developed strategies which enabled them to make a start by themselves and sustain their interest. They became systematic workers and produced more correct answers. More importantly, if correct solutions were not found they would persist by taking a different route. I was always greeted with much enthusiasm for the problem-solving session, even in the playground, as I got out of my car. One 8-year-old boy, Lucky, once asked me for my phone number so that he could phone me up and tell me how he 'was cracking the problem' we were working on. Cheeky, but very pleasing!

There is a notable absence of recent research literature in the UK about teaching problem solving. But from my experience, I would certainly recommend to teachers that they offer problem-solving sessions to pupils either in class or out of school hours. Before presenting some examples of problem-solving activities, I will draw your attention to what mathematics educators say about aspects of problem solving. Burton (1986) offers invaluable advice on aspects of problem solving and gives a collection of problem-solving ideas based on her research carried out with pupils aged 9 and 13. The teachers who worked with her were asked to spend an hour a week on problem solving. During that hour they were asked to change their role from one of responsibility for what the pupils do and learn to one of being a resource for pupils. They were encouraged not to give answers or methods but to provoke their pupils into searching for these themselves. This shift in roles had a profound effect on what children could do and understand. This kind of approach would provide an ideal way to challenge our able mathematicians.

Teachers will also find Ernest's (2000) list of problem-solving strategies useful. He maintains that some typical strategies that learners can use on a variety of complex problems and investigations are:

- representing the problem by drawing a diagram;
- trying to solve a simpler problem, in the hope that it will suggest a method;
- generating principles;
- making a table of results;
- putting the results in a helpful order;
- searching for a pattern among the data;

- thinking up a different approach and trying it out;
- checking or testing results.

These general strategies, Ernest maintains, are usually learnt by examples and are sometimes extended by the learner.

What I have discussed so far in the book should lead us to consider the following process:

$$\text{pupils} \xleftrightarrow[\text{challenges}]{\text{identification}} \text{problem solving}$$

Now I invite you to look at a problem-solving activity I have tried with children. These are selected examples which should provide some suggestions for teachers.

Problem 1 (Burton 1986)

There are some rabbits and some rabbit hutches. If seven rabbits are put in each rabbit hutch, one rabbit is left over. If nine rabbits are put in each hutch, one hutch is left empty. Can you find how many rabbit hutches and how many rabbits there are?

Initially many children were baffled by this problem and told me that they did not know where to start. Just by asking them to read the problem again and extract the information had been enough in some cases to get them started and find the correct solution. The children I worked with had not yet mastered algebra to be able to find a solution such as suggested by Burton:

Let h = number of hutches and r = number of rabbits

Then $r = 7h + 1$ and $r = 9 (h - 1)$

Therefore $7h + 1 = 9h - 9$

So $2h = 10, h = 5$ and $r = 36$

Some children found the solution to this problem (five hutches and 36 rabbits) by trial and error. Diagrams and models helped some of them to think about the problem and arrive at a solution. An example of how this problem was solved by a 9-year-old is included later in this chapter in the form of a diary entry. The joy of finding the solution had been remarkable on all occasions. Children also learned the important message that what may seem difficult or impossible can be solved by systematic thinking and persistence.

Select two problem-solving activities from a resource book and try to work them out. Consider:

- how will children tackle these?
- how would I go about helping them to solve these problems.

Encouraging children to use their metacognitive skills

Metacognition is a process which makes one aware of their own thinking processes and by reflecting on these become more effective as a thinker. Metacognition makes a person aware of their own condition and how they have acquired knowledge, and hence focus on the learning process. Flavell describes metacognition as:

> any knowledge or cognitive activity that takes as its object, or regulates, any aspect of any cognitive enterprise. It is called metacognition because its core meaning is cognition about cognition.   (1985: 104)

Research has shown that gifted children have higher levels of metacognitive skills and therefore should be given opportunities for improving these. Pupils who acquire metacognitive skills should become more effective in their thinking, reasoning and problem solving and so produce more thoughtful and creative work.

How do we encourage metacognitve processes in practical terms? Discussion and reflection during lessons helps. Self-assessment of one's learning must help too. I often ask all children to complete a grid to facilitate thinking about their learning (Figure 3.4).

One of the most effective ways in which I have involved pupils in thinking about what they had been doing was through the introduction of mathematical learning journals or diaries. Children can be encouraged to keep diaries for recording their mathematical thinking. When I first tried it, the initial reaction from the pupils was one of dismay because 'we don't keep diaries for mathematics; it's only for English'. Ten-year-old Darren was not familiar with a diary for mathematics. I explained that the purpose of the diary was to enable him to think mathematically as proper mathematicians do, keeping a record of what he is doing. Figure 3.5 is an example of an entry in a mathematical diary; it speaks for their effectiveness.

Achievement Record ☺

This week we have learnt about the topic:

Fractions

I think this topic is about: ~~thith things~~. dividing and sharing

A list of mathematical words and symbols I have learnt: equivalent, half, fractions, out of, quaters, third, three quaters, cut into bits.

A list of mathematical facts I have learnt: $\frac{1}{2} + \frac{1}{2} = 1$ whole two thirds and 1 third is One whole To find a third of Something I have to divide by three.

Make comments about how well you think you have learnt this topic and any other views you have:

I Learnt a lot about frations, I know how to work out equivalent fractions.
I like frations because when you do them you feel good. You feel happy you have figured it out. You don't have to say "Oh no! You can say 'I've finished'.

By: Chima Hogan
Age Nine and three quaters
School Archbishop Sumners School

Figure 3.4   Nina's completed 'thinking about learning' grid

The Rabbits and Hutches Problem.

The problem was: There are some rabbit hutches. If 7 rabbits are put in each hutch, 1 rabbit is left over. If 9 rabbits are put in each rabbit hutch, 1 hutch is left over.

Can you find out how many rabbit hutches and how many rabbits there are?

I solved it by going through my 7 times tables. Then I did the 9 times tables. I then added 1 to my 7 times tables. I picked 2 numbers, one from my 7 times table and one from my 9 times tables that are the same. For example:

$$7 + 1 = 8$$ 9
$$14 + 1 = 15$$ 18
$$21 + 1 = 22$$ 27   I then circled the 2
$$28 + 1 = 29$$ (36)   numbers then went
(5) →$$35 + 1 = (36)$$ 45   down the 7 times table
×7 $$42 + 1 = 43$$ 54   to find where I found
$$49 + 1 = 50$$ 63   the number that mat-
$$56 + 1 = 57$$ 72   ched the in the 9 times
$$63 + 1 = 64$$ 81   table.
$$70 + 1 = 71$$ 90
$$77 + 1 = 78.$$

re 36 rabbits and 5 hutches.   I found out there was

I enjoyed doing this. I felt great because I was the first to get it.

By Tammy Hogan 9 Archbishop Sumner School.

Figure 3.5   Tammy's diary entry

I am often asked how younger children, who cannot write, keep diaries. Diaries do not have to be written by the children; the teacher can record children's ideas and thoughts in a big class diary!

Providing special programmes: mathematics master classes

Mathematics master classes have been in existence for a very long time. Recently, these have been more frequently organised by individual schools, registered charities and universities. Since 1999, funding has been made available to LEAs by the DfEE for organising summer schools for the gifted and talented. Mathematics sessions feature in most summer school programmes. As a result, we have been getting an increasing number of requests at Brunel University for guidance on how to run an effective programme for pupils who attend these classes.

For those who may be interested in aspects of running effective mathematics enrichment programmes through master classes or summer schools, I will describe a programme I directed which had been commissioned by Wandsworth LEA. What I am about to share is based on evaluation and notes over a period of five years and involved over 600 11-year-old pupils. These pupils were selected on the basis of their SATs results and teachers' observations.

Children's perceptions

During the first day, before the tutors began any sort of work, all the pupils were asked to complete a questionnaire. The aim of the questionnaire was to find out pupils' perceptions of what mathematics was about, their attitudes to mathematics and what they thought of their achievements in mathematics. The purpose of asking children to complete the questionnaire was two-fold: first, it helped the tutors to make an initial assessment of aspects of the children's thinking about the subject; and second, for evaluation purposes by comparing the responses in the initial questionnaire with a similar questionnaire at the end of the programme. Tutors also kept samples of children's written work, their comments and copies of their written monthly evaluations. The observation from Sam (Figure 3.6) may be of interest to the readers of this book.

- Most children perceived mathematics as working with 'number' especially, doing 'sums'.
- About 75 per cent of the children said they liked mathematics. The most frequent reasons for liking the subject included getting correct answers

and getting certificates or some kind of public acknowledgement of their abilities. Those who articulated their dislike of the subject gave their reasons as 'boring' or 'too easy'.

- Ninety-two per cent of the children felt that they were 'good' at mathematics. Reasons included 'being in the top set or group' for mathematics lessons or being selected for the master classes. Those who did not consider they were not 'good' at the subject stated their reasons as 'I don't know if I am good' or 'no one has told me I am good at maths'.

Principles

Based on the responses to the questionnaires and our understanding of international research which focuses on the needs of mathematically able pupils, we set some principles for guidance. The following principles guided us in providing useful experiences to pupils.

- Children need to change their perception that mathematics is all about numbers and sums. There is a need to create a different mathematical environment to what they experienced in school.
- Able pupils are capable of learning advanced concepts in mathematics which are not usually covered in school. Thinking in depth about mathematical ideas, therefore, should be an important aim of such a programme.
- Regardless of the nature of the concepts dealt with, ideas should be presented in motivating contexts. This ruled out giving them books used by older pupils or distance learning packages which have not been designed for younger children.
- An element of competition was to be introduced. As the pupils were selected according to their higher ability this was not difficult to manage. However, some of the competitive work was organised in teams so that some individual children would not struggle with ideas. Group work also made pupils learn from each other and challenge the ideas of others.
- Children should be introduced to presentation formats which professionals use. For example, when conducting an investigation on aspects of 'Smarties' or 'Newspapers', it is necessary to teach them methods of presenting data. The use of spreadsheets and databases and other resources in ICT should be encouraged.
- Children should be provided with opportunities to develop mathematical processes. They should be familiar with the methodology of mathematics. The key concepts model for learning mathematics provides a basis for the design and selection of activities.

Sample of activities

The Gauss problem

It was felt that pupils who participated in special mathematics classes would benefit from learning about successful mathematicians from the past who have made significant contributions to the development of the subject. Mathematical behaviours which were adopted by mathematicians in the past are worthy of consideration. During the first session, pupils were introduced to Carl Friedrich Gauss (1777–1855), who was mentioned in Chapter 2, as one of the greatest mathematicians of all time. We talked about how Gauss used his mathematical ability by asking children to solve a problem similar to the one he solved when he was 7 years old.

Add all the whole numbers from 1 to 100

In most cases children started adding the numbers from 1 and writing subtotals. Many others added 1 to 10 and multiplied it by 10. Only when they were told it was not a good idea and, in any case, it would be very difficult to add all those numbers did we give them a strong clue that perhaps they should break it down in to a manageable problem. The idea based on pairing of the numbers was suggested by one pupil:

$$1 + 2 + 3 + 4 + 5 + \ldots + 95 + 96 + 97 + 98 + 99 + 100$$

They realised that there were 50 pairs which added up to 101, giving an answer of $50 \times 101 = 5050$.

Through this activity, we let the children see the elegance and the generalisable nature of the solution. Then the fun really started when the children wanted to try adding the numbers 1 to 50, then 1 to 200 , 1 to 1000 and so on. With some encouragement and prompts children realised that they could generalise the solution. The formula $n(n + 1)/2$ would always give them the correct totals for any set of numbers. They felt empowered that they would be able to use the generalisation when an opportunity arose. During the second day, many pupils told the tutors that they had been boasting about being able to add all the numbers from 1 to 1,000,000 within a few seconds! One boy proudly admitted that he told everyone he could find, including the people waiting at a bus-stop, about his newly learnt 'mathematical trick'. The fact that Gauss often worked on mathematical problems in his own time as school work did not always challenge him was an important message the pupils take with them at the end of the first day.

Spiderland

Most children who attended the programme were reasonably competent in performing the four operations. Many performed them by remembering rules which sometimes let them down. Many were unable to check the reasonableness of their answers; this was due to a lack understanding of the principles of place value of numbers. 'Spiderland' was offered to get children to think about the base 10 number system, being based on the grouping and regrouping concept. The activity 'mathematics in Spiderland' (Casey and Koshy 1995), which had been used in a mathematics enrichment project, encouraged children to reflect on the rules governing our number system while having fun.

The activity asked children to imagine that spiders used a number system based on groupings in eight, as they have eight legs, and to design some sums using all the operations, multiplication tables and a dot-to-dot picture which could possibly be used by spider children. The outputs were not only interesting, they also showed evidence of children developing an enhanced understanding of the base ten system, which in turn, showed an increase in their facility to reason and estimate while carrying out numerical work.

Many of the activities used in the programme were open-ended investigations which encouraged children to think of the underlying principles behind the mathematical ideas and encouraged them to be engaged in mathematical processes such as being systematic, hypothesising conjecturing, generalising, offering proof, finding counter arguments and so on.

Mathematical mastermind

Mathematical mastermind was always the highlight of the master classes programme. This was similar in nature to many national and international competitions, the only difference being that the competition was between teams of pupils. The emphasis was on group learning and problem solving. Questions were given out to groups. The challenges included facts such as names and terminology, knowledge of number bonds, a challenging hypothesis to examine, true and false questions and problem-solving activities. Tutors were impressed by the way children discussed ideas and learnt from each other and at the same time the competitive element added to their motivation and commitment.

Benefits

At the end of a series of masterclass sessions, children were asked to evaluate their learning and say what they felt about the programme. Some children were also interviewed. The comments made by 8-year-old Sam were a typical evaluation (Figure 3.6).

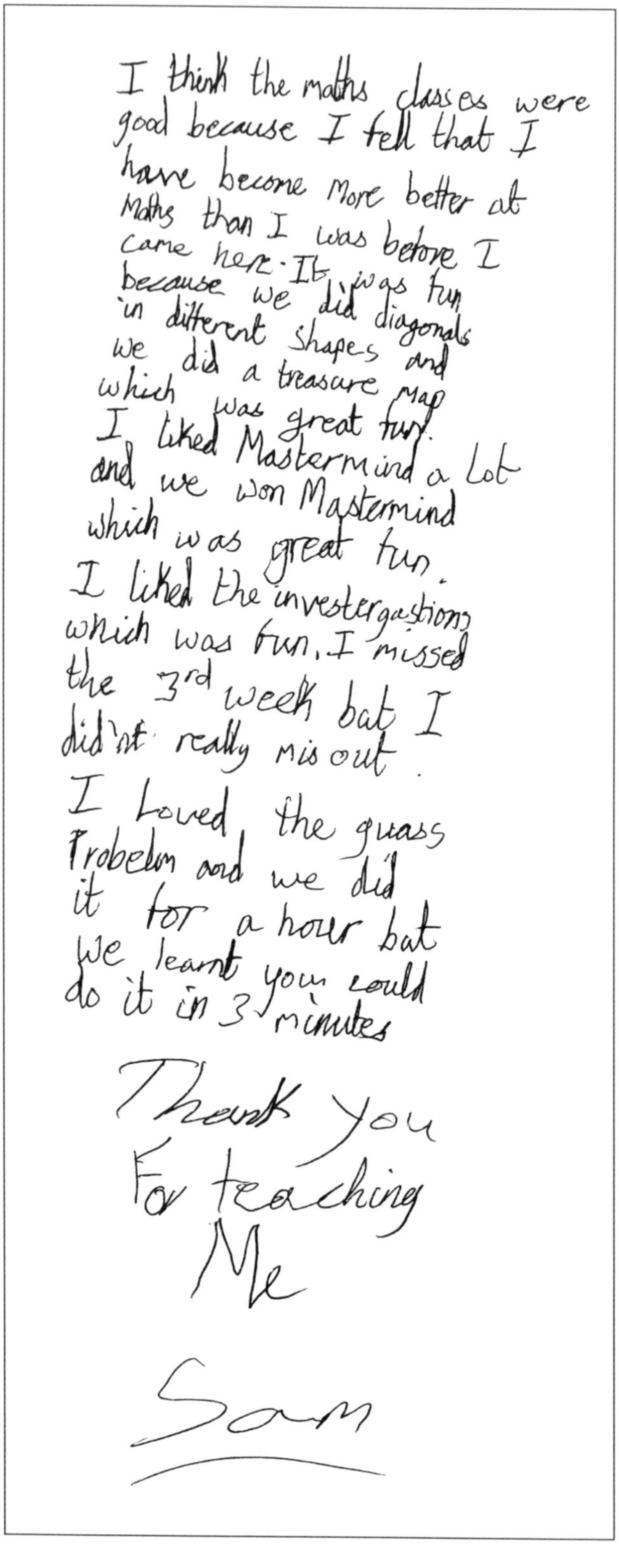

I think the maths classes were good because I felt that I have become more better at Maths than I was before I came here. It was fun because we did diagonals in different shapes and we did a treasure map which was great fun.
I liked Mastermind a lot and we won Mastermind which was great fun.
I liked the investergastions which was fun. I missed the 3rd week but I did'nt really mis out.
I loved the guass Trobelm and we did it for a hour but we learnt you could do it in 3 minutes

Thank you For teaching Me

Sam

Figure 3.6 Sam's comments

One of the features of our master classes was that we wanted teachers to be involved in all stages: in the selection of the children, the planning activities, and in being participant observers during the programme and sharing the evaluations.

Summary

You may notice that this is a very long chapter; in fact it is the longest in the book. This reflects my thinking that provision is the most important aspect when you consider the whole field of gifted education. Even effective identification, I hope I argued convincingly, is only possible and sensible if the provision is right.

In the first part of this chapter, I set out some general principles for making provision for mathematically able pupils. I introduced the key concepts model which was developed at Brunel University to provide a framework for provision. I provided a list of questions which should guide a practising teacher to select suitable activities for extending the able mathematicians in their care. The section on how to incorporate higher levels of thinking in the mathematics lessons was included, along with some guidance on problem solving and the setting up of master classes for mathematics.

CHAPTER 4

The National Numeracy Strategy and the able mathematician

The introduction in the UK of the National Numeracy Strategy in September 1999 has been the most influential initiative in mathematics education since the implementation of the National Curriculum in 1989. Most schools follow the recommendations of the National Numeracy Strategy to teach mathematics to pupils in Key Stages 1 and 2. A framework for teaching mathematics from Reception to Year 6 (hereafter called the Framework) was issued in 1999 (DfEE 1999b) and a further set of guidelines (DfEE 2000b) has been sent to schools for use with Year 7 pupils. The introduction of the National Numeracy Strategy will have a profound effect on the teaching and learning of mathematics in the UK and, naturally, it will have implications for teaching mathematically able pupils. So, I feel it is important that we look at the reasons for the introduction of the strategy, the nature of the guidance provided in the Framework and the structure of the recommended daily mathematics lesson within the context of teaching able pupils.

Before I start discussing aspects of the National Numeracy Strategy, I would like to draw readers' attention to the fact that the word 'numeracy' refers to all aspects of mathematics, and not just to skills in using number.

Why was the National Numeracy Strategy introduced?

The teaching of numeracy has caused concern for a number of years. Reports from OFSTED and international comparisons of the mathematical achievement of pupils, such as the Third International Mathematics and Science Survey (TIMSS) highlighted that all was not well with the numerical competence of children in the UK. This was one of the major reasons for the introduction of the National Numeracy Strategy. Awareness of the problems experienced by pupils in various aspects of numeracy is not new. I have described in more detail (Koshy 1999) how the concerns have been articulated by mathematics educators and government ministers since 1976, when James Callaghan expressed his unease arising from complaints from industry that new recruits did not possess the numerical tools to perform their duties. How does this relate to able pupils? If general standards in mathematics are lower in this country, it has implications for all children's performance, including that of the mathematically able. The findings of the international comparisons suggest that our most able mathematicians may also not be achieving their full potential. One significant question I would like to raise in this context is whether we have been setting high expectations for all pupils or whether we have been content to teach the minimalist requirements of the statutory National Curriculum? Although any answers to this question can only remain a matter for speculation, one sensible strategy to raise the level of mathematical achievement is to plan a teaching programme with the most able pupils in mind.

Able children and numeracy

In the *National Literacy and Numeracy Strategies: Guidance on Teaching Able Children* (DfEE 2000b), able pupils are described as being 'quick to understand' and 'apply their knowledge and skills in creative and original ways'. A list of characteristics of pupils who are particularly able in mathematics, provided in this document, is a good starting point for considering aspects of the most appropriate provision. This list says mathematically able pupils:

- have the ability to generalise patterns and relationships and approaches to problem solving;
- are persistent and flexible in their search for solutions
- are able to develop logical arguments, often taking valid shortcuts;
- are capable of using mathematical symbols confidently;
- are able to grasp new material rapidly;
- are not always exceptional in carrying out calculations, but may see calculations as detail and less important than the problem as a whole.

Readers may notice that the above list contains many of the features of mathematically able pupils based on research that I referred to in Chapter 2. This is reassuring and provides us with a framework for considering how aspects of the National Numeracy Strategy may enhance the opportunities for able pupils.

The definition of numeracy

The following definition is given by the National Numeracy Task Force:

> Numeracy at Key Stages 1 and 2 is a proficiency which involves a confidence and competence with numbers and measures. It requires an understanding of the number system, a repertoire of computational skills and an inclination and an ability to solve number problems in a variety of contexts. Numeracy also demands practical understanding of the ways in which information is gathered by counting and measuring, and is presented in graphs, diagrams, charts and tables.
> (DfEE 1998: 4)

All the objectives highlighted are absolutely vital for fulfilling mathematical promise. Mathematically able pupils, for example, need to develop confidence in their ability which emanates from their competence with mathematical skills and their understanding of mathematical concepts. Able mathematicians have the ability to be good problem solvers and we can see that the National Numeracy Strategy gives a high status to problem solving, estimating and reasoning skills. The intentions of the strategy, therefore, should enhance the learning opportunities for the mathematically able.

The purpose of issuing the Framework, guidance for teaching mathematics from Reception to Year 6, is to help schools to 'set appropriately high expectations for all pupils and understand how pupils should progress through the primary years'. This important message, again, underpins the good intentions of the National Numeracy Strategy and is of significant importance when we consider how it can affect the teaching and learning of mathematics in relation to able pupils. What are the positive features of the Framework which will support able pupils?

Mathematical content

Strengths
The Framework offers clear and specific guidance on the mathematical content which is to be taught; this also matches the requirements of the National Curriculum. It contains yearly programmes of what should be taught. Key objectives are provided and supplementary examples are given. There are several ways in which this is helpful for teachers in the context of providing for mathematically able pupils: let me share some of my observations:

- My work with many able pupils has shown that they do not always have the necessary mathematical knowledge, skills and conceptual understanding needed in order to perform at their best. Quite often pupils who have very high levels of problem-solving ability have been seen to flounder due to their lack of numerical proficiency and fluency. The structure of the Framework will enable teachers to monitor pupils' content coverage and performance and to ensure that there are no gaps in their understanding. Once the basic facts and skills are mastered – able pupils often learn these faster than less able pupils – they can be given more challenging tasks which require higher levels of thinking. These tasks, of course, will also help the children to consolidate their knowledge of facts and skills.

- Many primary teachers feel they do not possess adequate subject knowledge in order to extend able children. Most primary schools do not have mathematics subject specialists to teach the subject. In the past few months I have been told by many teachers that the explanations of mathematical content, the sequencing of ideas and examples given in the Framework have helped to ease their anxiety and lack of confidence in deciding what to teach and how to teach it. They also tell me that topics which they avoided teaching in the past due to their own lack of understanding are now being taught. This must help the mathematically able.

- The emphasis on the correct use of mathematical language and symbols is another aspect which supports the more effective teaching of mathematics.

- The list of key objectives provided in the Framework offers opportunities for continuously assessing children's knowledge and understanding of mathematical concepts. As assessment is the key to effective and fair identification, this aspect of the Framework is another welcome addition which should support teaching mathematics to able children.

Possible concerns

- It is useful to remember that some able pupils may already know much of the content the teacher plans to teach them. They may have learnt it from parents, brothers or sisters and, in some cases, may have taught themselves from books and other resources. This is not all that common, but if this is the case, being asked to learn what has already been mastered can lead to boredom and the development of negative attitudes. In these circumstances it will be necessary to streamline the content for that particular pupil and to set other work which can be done independently or with the support of a mentor. These strategies will be dealt with, in greater detail, in Chapter 5.

- Concern is also expressed by mathematics educators (Hughes *et al.* 2000) that as the National Numeracy Strategy is being implemented in schools, teachers may give priority to teaching the skills of numeracy and relatively little attention to teaching the application of these skills. If this becomes the case, able pupils who possess high levels of investigative skills could be at a disadvantage. Let us hope that the problem-solving strand of the numeracy Framework will ensure coverage of the using and applying aspects of mathematics.

Aspects of calculations

Strengths
The Framework provides extensive guidance on aspects of calculation. Mental and oral mathematics are given a high profile. The Framework encourages estimating and checking skills to assess the 'reasonableness' of solutions. Flexibility in the use of algorithms is another feature. How do these support the able mathematicians? The following observations may be of interest:

- All children, including able pupils, need to develop effective strategies for mental calculations. It cannot be assumed that an able child will automatically learn and be able to recall aspects of mental calculations. Fluency and speed in calculations can greatly assist an able child to take on challenging projects without having to struggle with arithmetic calculations.
- Oral mathematics and the opportunities to explain ideas and methods help all children to learn new ideas and to refine existing ones. Able pupils often benefit from discussion and questioning which challenge their ideas and create a *cognitive conflict* which promotes greater understanding of mathematical concepts. This includes both discussions with the teacher and with other pupils.
- All children need to develop their effective and efficient algorithms, ones which, as Casey (2000) says, 'do the job better than other algorithms'. The flexibility of algorithms encouraged by the Framework would enable able pupils to think independently and offer creative solutions.
- The Framework discourages the limitations often put on younger pupils dealing with only smaller numbers before they are considered 'ready' to encounter larger numbers. This offers better scope for able pupils to explore larger numbers.
- The idea of introducing algebraic ideas from an early stage, recommended by the Framework, helps pupils to develop the skills of forming and solving equations, using inverses, identifying patterns and expressing relationships. As able pupils possess a greater degree of capability to use these processes, it opens up possibilities for teachers to provide them with more challenging activities which involve those processes.

Points of concern

- Due to the wide range of ability of children, especially as they get older, it may become difficult to manage differentiation within a mental and oral lesson on every occasion. In spite of using different strategies, teachers often say that they feel frustrated when they have to ignore some very bright pupils who have their hands up on every occasion. Able pupils, similarly, may feel that their contributions are not important or valued. Although the numeracy training sessions and resources offer useful strategies for differentiating the mental and oral starter lessons, this is an area which still causes concern to many teachers.

Teaching strategies

Strengths

- The Framework recommends that each teacher provides a daily mathematics lesson which should last about 45 minutes in Key Stage 1 and between 50 and 60 minutes in Key Stage 2. Most of this time is to be used for direct teaching and questioning. A list of features of 'good direct teaching' is also given in the Framework which includes:
  - Directing
  - Instructing
  - Demonstrating
  - Explaining and illustrating
  - Questioning and discussing
  - Consolidating
  - Evaluating pupils' responses
  - Summarising
  These aspects of good practice will contribute to the quality of teaching and learning for all pupils, including able pupils.
- The structure of the mathematics lesson helps the teacher to keep a good pace. I have observed mathematics lessons in the past which were slow in pace and which put able pupils at a disadvantage. In many instances, when they finished the assigned work faster, they were either asked to do more of the same type of work or to 'get on with something else'. Sometimes they were even asked to 'help' other children to catch up with the rest of the class.
- The listed elements of good 'direct teaching' constitute good practice in teaching mathematics and should lead to more effective learning and retention of ideas.

List of concerns

- As with all good intentions, the question of pace is also a cause for concern for many teachers. They feel guilty about leaving some children behind, especially those children who may have special needs or have language problems. Many schools are now setting children, which enables the teacher to keep a good pace, as the ability range in a 'set' group is likely to be narrower. Aspects of setting are discussed in Chapter 5.
- Although the well paced style of instruction has advantages for the able pupil, there is concern that rigid time constraints may kill off the enthusiasm of pupils who may want to spend more time on a line of enquiry. It should be remembered that the 'timed' lesson is only a guide. Care should be taken to ensure that able pupils are given other opportunities for extended work; this could be done by letting them continue their investigations for more than one lesson, setting a challenge for homework or by allowing time in the day to pursue their own enquiries.
- The other concern is the degree to which the elements of good direct teaching are applicable to able pupils if they are taught in mixed ability groups. While questioning, explaining, discussing and evaluating are useful processes to develop in all pupils, I feel that bright pupils are unlikely to need the same amount of direction and instruction as some of their peer groups. Again, some kind of grouping or setting may be necessary to address this.

The three-part daily mathematics lesson

One of the recommendations of the National Numearcy Strategy is that all schools should follow a structured daily lesson. In this section we will look at ways in which the needs of mathematically able pupils can be met within that structure. The format of the daily lesson provided by the Framework is as follows:

- *Oral work and mental calculation* (about 5 to 10 minutes)
  Whole-class work to rehearse, sharpen and develop mental and oral skills.

- *The main teaching activity* (about 30 to 40 minutes)
  Teaching input and pupil activities.Work as a whole class, in groups, in pairs or individuals

- *A plenary to round off the lesson* (about 10 to 15 minutes)
  Work with the whole class to sort out misconceptions and identify progress, to summarise key ideas and what to remember, to make links to other work and discuss the next steps, and to set work to do at home.

Providing for able pupils in the daily mathematics lesson

In this section we will explore some useful strategies to maximise the learning opportunities for able pupils. I will also provide some examples of activities I have observed in classrooms. The examples were selected on the basis of two factors: first, where the teachers concerned made a positive effort to meet the needs of able pupils; and second, where I felt that pupils were suitably challenged and showed much enthusiasm. Readers will find the books of Collins Mental Maths series (Resources section) useful for stimulating ideas and for strategies for differentiation.

Mental and oral starter

It is very likely that the 30 or more children in a class are at different stages of development. The spread of ability in the mathematics classroom can be very wide: the Cockcroft Report (Cockcroft 1982) claims up to a 'seven year difference'. Within that range we need to think of strategies to achieve differentiation. Here are some suggestions:

- Do not always start with easy questions which will inevitably be even easier for able pupils and will result in their hands being ignored all the time. Sometimes, it is good to start with a challenging question targeting able pupils and encouraging them to explain their method carefully. This will not only acknowledge their ability, but will help other children listen and take note of newer and perhaps more effective strategies.
- Choose some open-ended questions and encourage able pupils to respond at their own level of ability. For example, instead of starting with a problem such as 'add 24 + 38 in your head', phrase the problem as 'I added two numbers and got the answer 62. What were my two numbers?' Accept all correct answers and encourage pupils to explain their method. In a 'percentage' lesson, instead of always asking questions which involve finding an answer ('What is 25 per cent of £30?'), try 'A shop asked me to pay a 25 per cent deposit to buy a bicycle. If I paid £7.50, how much does the bicycle cost?' This type of question encourages a higher level of thinking.

- Instead of having all the children facing the teacher answering questions which involve one-step calculations all the time, encourage them to work in small groups or pairs to generate responses. This will enable pupils of the same ability to work together.
- When responding to questions, encourage children to record their answers on cards and to hold them up when asked. This enables teachers to skim through the answers quickly before selecting pupils to share their answers.

Examples from the classroom

Eileen's lesson

One day a week, Eileen's children work in small groups for their mental starters. She introduced a 'What is my secret number?' activity to the whole class of 6-year-olds, using an example of medium difficulty. She asked them to work out the mystery number from the clues:

> I am thinking of a number.
>
> It is lower than 40.
>
> It is higher than 26.
>
> If you count in 5s you will land on that number.
>
> It is an odd number.

After the children had figured out the mystery number (35) she asks them to work in smaller groups, working out other mystery numbers on worksheets. The questions were differentiated in order to match their capability. The following question was among those for the 'able' group:

> The mystery number is odd.
>
> It is a multiple of 3 and 7.
>
> It is between 60 and 100.

Mike's lesson

Mike is a Year 3 teacher. His weekly plan for the oral and mental starter highlighted 'addition strategies'. He started the lesson explaining an alphabet code which assigns a value for each letter (Figure 4.1). Mike asked his children to calculate the following questions using the alphabet code:

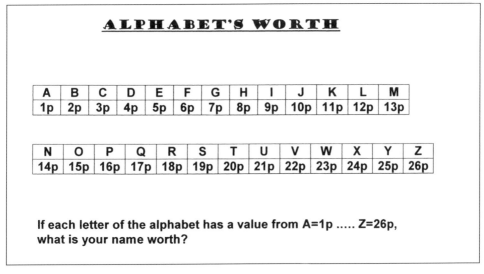

ALPHABET'S WORTH

A	B	C	D	E	F	G	H	I	J	K	L	M
1p	2p	3p	4p	5p	6p	7p	8p	9p	10p	11p	12p	13p

N	O	P	Q	R	S	T	U	V	W	X	Y	Z
14p	15p	16p	17p	18p	19p	20p	21p	22p	23p	24p	25p	26p

If each letter of the alphabet has a value from A=1p ..... Z=26p, what is your name worth?

Figure 4.1   Mike's alphabet code

- 'How much is an apple worth?'
- 'What is an ant worth?'
- 'Which is worth more, a bee or an ant?'
- 'Think of the most "valuable" three-letter word, in one minute.'

Children were asked to explain and share their strategies for adding a string of numbers. The process of estimating was also highlighted in this lesson.

Then Mike asked children to investigate more complex questions with their partners. he gave out differentiated questions and pencils to jot down ideas on different coloured papers. The 'able' pairs had the following questions on their 'yellow' sheet:

- 'Find the most valuable "animal" word starting with the letter B.'
- 'Find one word which is worth exactly £1.00.'
- 'Find the name of a month worth 94p.'
- 'Find the name of a food item which is between 80 and 110 pence.'

Barbara's lesson

Barbara has a Year 6 class. The mental starters she plans involve a range of styles. One lesson a week is dedicated to children exploring each others' mental strategies. Children are asked to work in pairs to do mental calculations. Each pair is given three questions to ask each other.

For example, Lena and David were given a sheet with the instructions:

Take turns and ask your partner to do the following calculations mentally. No writing is allowed on this occasion. After your partner has completed the calculations and given you answers, find out how he or she worked them out. Find out as much as you can by asking them questions and encouraging them to talk.

During the same lesson, four pairs were given the task sheet in Figure 4.2, while the rest of the class were working in pairs on calculations which were less demanding. On this occasion the teacher discussed the tasks in two groups separately; the more able in one group and the rest of the class in the other. This enabled Barbara to pitch the work at a higher level for her able pupils.

A's Card	B's Card
162 + 79	268 + 78
136 − 17	237 − 18
231 × 17	311 × 19
225 ÷ 25	231 ÷ 30

Figure 4.2   Barbara's task sheet for the more able pupils

Dick's class

Dick's Year 7 class has a very wide range of ability. Although pupils are set for mathematics, differentiation is still a challenging issue within his middle group. During one of his mental mathematics sessions, he explained the task: 'Search for the signs'. They must study the sum carefully and using their knowledge about number operations, fill in the missing signs. Afterwards, he provided the children with sets of sums with missing signs (Figure 4.3).

On this occasion, instead of bringing all the children back for a whole class discussion, he asked them to conduct a discussion within their groups. During such sessions he encourages different pupils to take on the role of the teacher. This, according to Dick, works 'extremely well' and 'it is good for the pupils to take the responsibility of asking the questions rather than answer the teachers' questions all the time'.

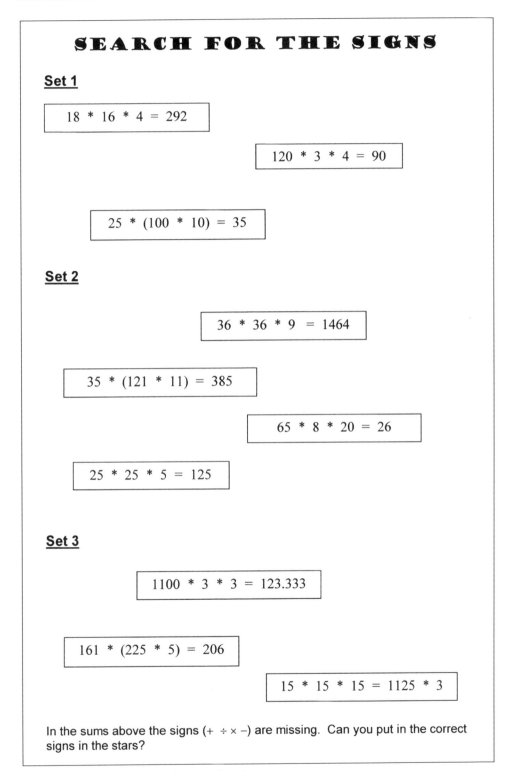

SEARCH FOR THE SIGNS

Set 1

18 * 16 * 4 = 292

120 * 3 * 4 = 90

25 * (100 * 10) = 35

Set 2

36 * 36 * 9 = 1464

35 * (121 * 11) = 385

65 * 8 * 20 = 26

25 * 25 * 5 = 125

Set 3

1100 * 3 * 3 = 123.333

161 * (225 * 5) = 206

15 * 15 * 15 = 1125 * 3

In the sums above the signs (+ ÷ × −) are missing. Can you put in the correct signs in the stars?

Figure 4.3   Dick's 'Search for the signs'

Jane's class

Jane has a class of Year 4 children, who 'love doing number puzzles'. Jane is convinced that number puzzles encourage children to develop a feel for number, fluency, estimating and grasping skills. During one of the mental mathematics sessions, she gave the children the following puzzle shown in Figure 4.4.

She labels the puzzles 'easy' 'hard', 'very hard' and 'impossible', and asks the children to choose which ones they want to do. She was surprised that many children who she had not expected to select the 'hard 'and 'very hard' sums did so.

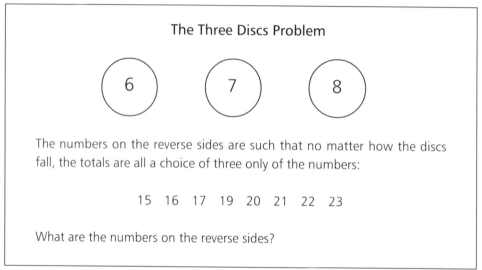

The Three Discs Problem

6   7   8

The numbers on the reverse sides are such that no matter how the discs fall, the totals are all a choice of three only of the numbers:

15   16   17   19   20   21   22   23

What are the numbers on the reverse sides?

Figure 4.4   Jane's 'Three discs problem'

The main part of the lesson

Whether or not the class is set or taught in mixed ability groups, much of the differentiated provision in a lesson can be achieved in this part of the daily lesson. During this part of the lesson much of the work can be done in smaller groups. It is also possible for children who are exceptionally able to pursue work independently or perhaps work with a mentor on more advanced concepts for part of the time. The following strategies are worthy of consideration:

- When new work is introduced, it is worth remembering that some pupils may already know what they are going to be asked to do. In this case, some curriculum compacting will be necessary. Curriculum compacting or streamlining the content would involve spending less time on explaining ideas which have already been mastered and avoiding undue repetition and reinforcement. It is important to remember that able pupils do not need to start on the same page as everyone else; they can miss earlier stages of a topic or a task. Many schools are, at present, buying 'ready-made' lesson plans for use. Using these lessons, designed for 'average' pupils without adjustment, could lead to boredom and frustration for the most able mathematicians.

- It will be necessary to employ differentiation by task and by outcome. For example, when most of the class is working on place value of whole numbers, the 'able' group can be introduced to decimal notation and operations involving decimals or they can be introduced to other related concepts which may not be within the grasp of other pupils in the class. We must resist giving children extra pages of the same kind of work!

- The main part of the lesson is an ideal time for able pupils to be engaged in investigations and open-ended mathematical enquiries. During this time they can pursue individual lines of enquiry. This will provide opportunities for in-depth exploration of mathematical ideas.

- Mathematically able pupils are capable of working for an extended period of time on a task; it is not necessary or advisable to conclude every task after one lesson. Encourage children to discuss the task with you to clarify their thinking and share their ideas before taking the ideas further, perhaps in the next session.

The key concepts model for provision described in Chapter 3 can be used as a guide in selecting a mathematics programme for the pupil. Revisiting the model will help us establish some principles for selecting activities for able pupils. First, it cannot be assumed that all mathematically able pupils have a sound knowledge base. They may learn fast, but knowing *facts* and having a mastery of *skills* and *fluency* will depend on what opportunities have been offered. Any activities which target mathematical facts, skills and the development of fluency, therefore, should be included in the programme; without these children cannot undertake investigative work efficiently. If pupils have a sufficient knowledge base, then the work set for them should aim for an in-depth learning of ideas. Pupils should be allowed time for exploring aspects of *algorithms* and the principles underlying them. Activities should appeal to their curiosity and enable them to exercise their creativity. Tasks should encourage them to look for patterns, hypothesise and conjecture and offer proof. Pupils should be given opportunities to detect *isomorphism*: similarities in mathematical structure. The activities described in the following sections have been selected with the key concepts model in mind.

Examples of main lesson activities from the classrooms

Consecutive sums

A class of Year 3 children was introduced to the activity on Thursday morning when the class usually investigates a mathematics idea. The teacher asked the children if they knew what consecutive numbers were. Having established that they are numbers next to each other on number line – a 'sort of number neighbours', as one child put it – the whole class listened to the teacher:

> What numbers can you make by adding consecutive numbers?
>
> $1 + 2 = 3$
>
> $2 + 3 = 5$
>
> Can you make 4 by adding consecutive numbers?

They established that it cannot be done.

> Try and find the numbers from 1 to 20 that can be made by adding consecutive numbers.

'Can we add more than two consecutive numbers to make a total?' asked Gary. 'Like $4 + 5 + 6$ making 15, which can also be made by adding 7 and 8', he explained.

Children were asked to work in groups and given three strips of paper and thick coloured felt-tip pens to record their discoveries, which they had to present to the rest of the class during the plenary session.

The whole class worked on this investigation for 25 minutes and established that it is not possible to make numbers 1, 2, 4, 8 and 16 by adding consecutive numbers.

Their hypotheses included :

- All odd numbers except one can be made by adding consecutive numbers. Why may this be?
- Some numbers can be made by adding two or three consecutive numbers.
- 6, 9 ,12 and 15 can be made by adding three consecutive numbers.
- 4, 8, and 16 cannot be made by adding three consecutive numbers. They are 'doubling'. Why?

The teacher suggested a new line of enquiry to those who had completed the initial task.

What do you notice about the three consecutive numbers which aᴑ multiples of 3?

Children discovered that the middle number is the total divided by tι Is it always the case? They tried some bigger numbers. It worked. The hypo esis seemed to work, supported by their examples.

The more able mathematicians were encouraged to put forward more conjectures.

What about numbers which can be made by adding five consecutive numbers? Are they multiples of 5? Try seven consecutive numbers.

The teacher then posed another question to the more able group.

I added five consecutive numbers and got 125 as the total. What were the consecutive numbers?

They soon worked out that 125 divided by 5 will give you the middle number, 25. So if you add 23, 24, 25, 26 and 27, you will get 125. The children soon established that this rule works for 7, 9 and so on. At the end of the lesson, a new question was put to the able group.

Can you think of a way of finding out which four consecutive numbers will add up to 46?

46 cannot be divided by 4 without a remainder. This was set as a challenge for homework to be recorded in their mathematics journals.

The above activity has some features which make it suitable for extending able pupils. Some of the features the activity offers are:

- open-ended possibilities – new conjectures can be made and explored at every stage;
- opportunities to have a common starting point, so that the investigations do not become a 'special' treat for some pupils;
- a context for meaningful mathematics recording;
- a context for using and applying mathematical operations.

In fact, in the school where this activity was observed, the consecutive number project lasted for five weeks for a group of able pupils. Their lively diaries of notes and discoveries were put up on the mathematics display table.

Elmer's buns

This activity was used by a Year 2 teacher. Her objectives for the lesson were problem solving, addition of a string of numbers, children posing their own problems and exploring strategies and recording them. The activity was presented to the whole class, as it is on the worksheet in Figure 4.5.

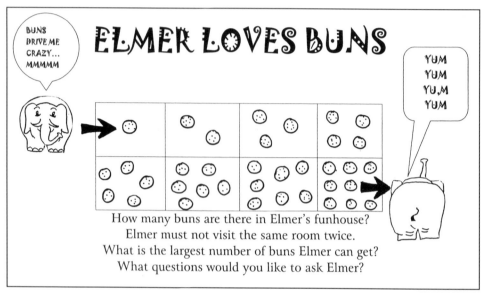

Figure 4.5 Elmer's buns

All the children were highly motivated by this activity. As in the case of the consecutive numbers activity, it provided opportunities for further investigation. The class teacher encouraged the children to ask their own questions and investigate them. New questions generated using the 'what if' strategy were:

- What if the shape of the house was different, say a square?
- What if the buns were arranged differently?
- How many different ways can Elmer go in and out of the house without visiting the same room twice?
- If each bun costs 23p, how much will it cost to buy the buns?

Investigating number patterns and sequences

Mathematics is a study of patterns. A good mathematician will find patterns and make sense of ideas and fit them into elegant generalisations. Mathematically able pupils are especially capable of this; the mathematics curriculum must provide them with opportunities to do this. The task in Figure 4.6,

What is the rule?

Use a table as shown to work out the rules of these sequences. Then complete the missing spaces

Sequence	20th term	50th term	Rule
4, 5, 6, 7, _, _, _	23	53	$t + 3$
23, 30, 37, 44, _, _, _	156	366	$(7 \times t) + 16$
8, 13, 18, 23, _, _, _	_	253	_
6, 10, 14, 18, _, _, _	_	_	_
1, 4, 7, 10, _, _, __	_	_	_
8, 22, 36, 50, _, _, _	274	_	_
4, 9, 14, 19, _, _, 34	_	_	_

Figure 4.6 Investigating number sequences

involving finding sequences, was differentiated before groups of Year 5 pupils were asked to work on it. Two groups were asked to complete the sequences, filling in the next three numbers in each sequence. The middle group was asked to fill in the missing numbers and work out what the 20th term would be. The group of able mathematicians was given the extension task:

- Fill in the missing numbers
- What is the 20th term?
- Can you find the rule to work out the 50th term?
- Test your rule to see if you can convince me (teacher) that your theory works

Expert estimators
Joseph teaches the Year 4 top set. The children are often told by their teacher that being a mathematician means that you will be able to estimate solutions and judge the reasonableness of those solutions. Using the 'Likely stories' idea (Figure 4.7) is one way of developing the skills of estimating and reasoning.

Likely Stories by Rebecca Parto
65 - St. Stephen's School

Using your mathematical talent, work out which of the following
could be likely or unlikely. Write either 'L' or 'U' next to it and give a
reason for your decision.

You have been alive for more than 4 million minutes ..U...... *If there are
60 minutes in an hour, 24 hours in a day, 28 days in
a month, 12 months in year, and you are 11 years old
than the amount of minutes is 5322240, add a little
more because most months have more than 28 days.*

Your weight is less than 2Kg...U...

*Even a baby weighs more than 2 kg. At the moment,
in relation to that, I weigh 28-29 kg, over 4 stone., as
26 kg is roughly 4 stone.*

A person's height can be 176 cm..L..

*A person can be around 2 metres= 200 cm, so it
is quite likely they could be 1m 76.*

8 dozen is less than 36...U..

*This is unlikely because I know 8 × 12 = 96, much
higher than 36. However it would depend how much
the 8 was worth. If it was 0.8 × 12 = 9.6, much less
than 36.*

A car can travel more than 1000 kilometres per minute...............

*A minute is very short, and at this moment of
technology, we haven't made a car as fast as that.
However in the near future, we could possibly invent one!*

Figure 4.7 Rebecca's 'Likely stories'

Joseph also encourages his children to construct true/false type questions
(Figure 4.8) for their 'mathematics challenge box'.

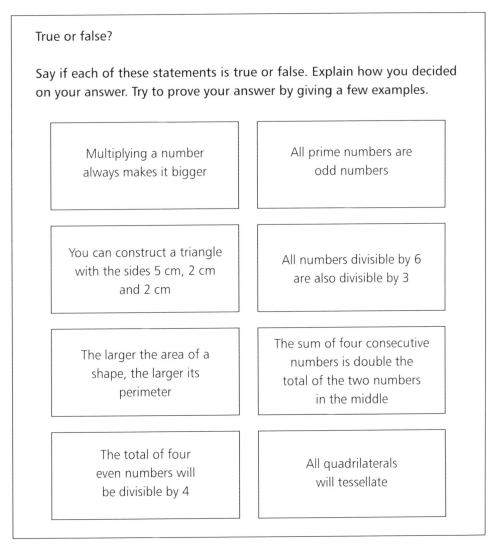

Figure 4.8 True or false?

Strive for the highest
A group of Year 7 pupils were given the game in Figure 4.9 by their teacher, whose listed objectives were: estimating skills, using a calculator correctly to find percentages and to be strategic thinkers.

Connect the percentages
When classes are set for mathematics, teachers will still have to differentiate the work for different groups. The activity in Figure 4.10 was offered to a top set of Year 6 children while the other sets worked with the same ideas using only whole numbers. The pupils were challenged and motivated.

Strive for the highest

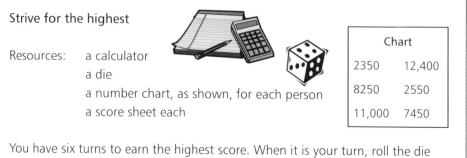

Resources: a calculator
a die
a number chart, as shown, for each person
a score sheet each

Chart	
2350	12,400
8250	2550
11,000	7450

You have six turns to earn the highest score. When it is your turn, roll the die and circle a number on the chart. This number cannot be used again. To calculate your score, increase the circled number by the percentage rolled on the die. For example, if you circled 2350 and rolled 4 on the die, your score is 2350 × 4% = 2444. Keep rolling the die and choosing a number. After six turns, add all your scores. The person with the highest score wins. You may decide to play again to improve your strategy.

- What advice would you give to a person who is going to play the game? Use your experience to guide you.
- What is the largest possible score you can get after six turns, if you were lucky? What conditions would be necessary for this to happen?

Figure 4.9 Strive for the highest

Connect the percentages

Take turns to choose a percentage and a number. Work out the result and cover that number on the grid. Your partner will check it on a calculator before you can claim that space.

Percentage	Number
40%	60
20%	35
30%	900
25%	88
90%	150

When you connect a line of four numbers, up, down or diagonally, you win a point. Keep playing until you have scored three points.

12	24	37.5	26.4	360	7
30	360	31.5	54	45	24
135	79.2	24	30	22	225
810	180	10.5	18	35.2	360
14	35.2	60	15	8.75	270

Figure 4.10 Connect the percentages

Plenary

Achieving differentiation is a challenge during all three stages of the daily mathematics lesson. Some teachers find plenaries the hardest to manage. It is perhaps hardly surprising as children learn at different rates and the levels of understanding of the concepts targeted during the lesson could extend over a wide range. This can make whole-class discussion of methods and outcomes difficult. It is even harder to have whole-class discussions of what has been learnt if the tasks set were different or when the 'able' group worked on a very complex extension to the main activity. So, what strategies can be used?

- Some plenaries can be conducted with smaller groups. This is particularly useful if the given tasks are different or when tasks have been pitched at a very complex level for able pupils.
- Remember that much learning can take place during plenaries when the able mathematicians are invited to make presentations of what they have learnt. This will make them feel that their contributions are valued.
- Able children can be invited to display the work done by the class. They can provide titles and captions, as well as explanations of mathematical ideas, to enhance the display.
- In an activity such as investigating number sequences, the able group can share their generalisations with the rest of the class.
- A mathematically able child can be asked to make up a glossary of mathematical words and symbols learnt during the lesson.
- Able pupils may be asked to prepare a quiz, based on what has been learnt.

Exceptionally able mathematicians

In the previous sections we dealt with a number of strategies and examples of work used by class teachers to extend the more able mathematicians. If you have an exceptionally able mathematician in your class, the extension activities suggested in this chapter may still not offer sufficient challenge and stimulation. In that case, other arrangements need to be made. Moving children up a year or seeking the help of a mentor who is a mathematics specialist and has an interest in helping young mathematicians needs to be explored. These flexible types of organisational styles are discussed in the next chapter.

Summary

In this chapter we looked at aspects of provision for pupils with high mathematical ability in the context of the National Numeracy Strategy. Aspects of the strategy which will undoubtedly support the able pupil were discussed; we also considered aspects which need extra caution. For the able mathematician the daily mathematics lesson (what many refer to as the numeracy hour) could be anything from a period of 'sad frustration' by being asked to learn facts already learnt and skills which have already been mastered to a 'happy hour' of knowledge exploration. It is the sensitivity of the teacher to the educational and emotional needs of the able child, as well as the ingenuity of questioning, which will determine the positioning of the child's experience within the spectrum of possibilities of the daily mathematics lesson.

CHAPTER 5

Organisation for teaching and learning mathematics

In this chapter we will explore aspects relating to the organisation of teaching and learning mathematics. The crucial factor in selecting any organisational style is that it must provide opportunities for all learners to maximise their learning; able pupils are included in this. The organisation of the school, how the classroom teaching is organised and the teaching styles adopted by the teacher will all have an impact on pupils' learning. This chapter deals with a number of organisational issues which are discussed in the context of provision for promising mathematicians. The starting point for such a discussion is to ask the question: what are the critical factors which may impact on the mathematical achievement of children? Take a few minutes to study the following list and then ask yourselves how these factors affect the learning of able mathematicians and then consider what is happening in your situation with regard to each of these.

Critical factors which influence learning mathematics.

A careful study of the literature of mathematics education and gifted education will highlight the following factors as of paramount importance:

- assessment of ability
- teaching styles
- organisational structures
- differentiation.

Each of these will now be considered.

The assessment of ability

At the time of the introduction of the National Curriculum and Statutory Guidelines in England and Wales, the following guidance was provided by the DES:

> The main objective of assessment arrangements will be to ensure that each pupil's attainment in a subject and the elements within it can be clearly identified and the results used to help the pupil's progress. It is essential that the assessment arrangements establish what children know, understand and can do in order that teachers and parents can identify their children's strengths and weaknesses, and plan the next step in their education.
>
> (DES Circular 5/89)

In the context of the classroom, the model shown in Figure 5.1, provided by Mitchell and Koshy (1995) shows the links which make learning effective.

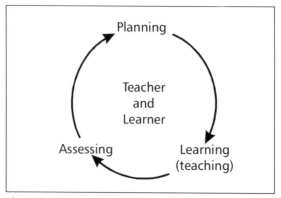

Figure 5.1 The planning, teaching, learning cycle (Mitchell and Koshy 1995)

Let us consider this model in the context of the able pupil. The teacher gathers information about the child's learning from a number of sources. This information is used in planning activities which match the ability of the child. Appropriate styles of teaching and organisational arrangements are adopted when teaching the pupil. When an effective match of content and teaching styles is employed the assessment of ability and of potential talent become more effective.

The assessment of abilities is a complex process. When developing assessment strategies for able pupils the following are worth remembering:

- Employ a range of assessment methods.
- Make sure that parents, teachers and pupils are involved in the assessment process and that each understands both the methodology used and the basis on which judgements are made.
- Use the information gathered from multiple sources to evaluate how effectively you are matching the curriculum to the needs of the pupil. This means considering how effective you are in extending the pupils' talents by planning a challenging, stimulating and enriching set of activities.

Ways of assessing pupils' mathematical achievement

In Chapter 2 we considered a range of useful strategies for identifying mathematically able pupils. In the following section I will share a list of strategies I offer during in-service courses. Many teachers have found these useful during parents' evenings, OFSTED inspections and for setting up Individual Learning Plans or setting targets for able pupils.

Everyday observations

These happen during normal mathematics lessons, by observing a child or listening to group discussions during work on an activity. Comments made by the child during mental and oral lessons or plenaries are particularly useful. These observations may be recorded as notes or anecdotes which will build up a true picture of the nature and extent of a pupil's ability.

The analysis of written work

This will involve assessing a pupil's written work. The product will be assessed against the learning intentions, processes and skills involved and any special talents demonstrated. Written products, diagrams and photographs of any models constructed can be stored as records or evidence of special skills and talents.

Gathering information through interviews

In order to gain a good understanding of a pupil's capability, it is sometimes useful to arrange an interview with the child when targeted questions can be asked. Asking pupils and parents to complete separate questionnaires about children's work habits can also provide valuable insights into the ways of their thinking and the nature of the strategies used. These can be logged in the pupils' records.

Pupils' self-assessment

Self-assessment encourages able pupils to reflect on their own learning and achievement as well as evaluate their methodology. Self-assessment will develop the pupil's metacognitive abilities. Able pupils have enhanced powers of metacognition. The following strategies can be used to encourage them to use their metacognitive abilities:

- Ask pupils to keep a mathematical journal. An example was included in Chapter 3. In the journals the pupils will chart their thinking processes, new ideas and developments and wrong directions as well as discoveries, generalisations, proofs and evaluations of what has been learnt and how the ideas were learnt.
- Encourage pupils to keep a portfolio of their best work and records of mathematical achievement. Details of competitions in which they have taken part, certificates and commendations won and examples of 'favourite' puzzles can also be kept in the portfolio.
- Ask pupils to create glossaries, fact books, self-designed games, problem-solving activities, interesting biographical details of well-known mathematicians and examples of number systems used in other countries.

Teaching styles

The mathematical tripod

Continuous teacher assessment provides information about children's learning; it also acts as an evaluative mechanism for teachers to assess the suitability of content and teaching styles. There is extensive literature dealing with teaching styles. Here, however, my focus is on the appropriateness of different teaching styles and methods for maximising the learning opportunities of able mathematicians. So what is an appropriate style of teaching the able? To answer that question, I will use a metaphor of a tripod, which keeps a camera still and steady so that it can take clear and high quality pictures. The three legs have to be stable and balanced for the camera (the child) to focus sharply and enhance the quality of the picture, which is the world of mathematics. The picture here represents all aspects of mathematics learning: facts, skills, conceptual understanding, problem-solving strategies and positive attitudes. The three legs of the tripod (Figure 5.2) represent three kinds of teaching styles; all have their particular role to play.

I will now describe the three types of teaching style which enable learning as described by the terms *osmosis*, *parenthesis* and *metamorphosis*. Although I

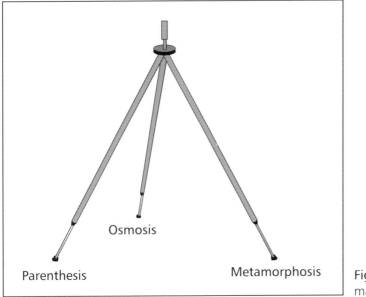

Osmosis

Parenthesis   Metamorphosis

Figure 5.2 The mathematical tripod

am describing the three modes separately, all three are employed by most teachers. It is the balance between them which is the key to selecting the most suitable style of teaching to be dominant for a particular group of pupils.

The teaching style which is symbolised by the term *osmosis* suggests the passage of knowledge from the teacher to the children. In science osmosis suggests the movement of a liquid from a higher concentration to an area of lower concentration. It is a one-way process through a membrane: the teacher/ pupil interface. This particular style is often referred to as the *transmission* model of teaching, with its behaviourist roots. Within mathematics, this mode will stress the acquisition of facts and a step-by-step mastery of skills. The teacher will tell and the children will listen and probably remember much of what has been told. In this passive mode of learning there will be little discussion, sharing of ideas or questioning. After the instruction phase, pages from the textbooks may be given to the pupils to judge the extent to which they have mastered what has been taught. The outcomes are easy to test against a set of objectives. Those who advocate testing and awarding grades relating to a fixed criteria of objectives may find this an acceptable mode of teaching.

How appropriate is this model of learning for the able learner? There is general agreement that gifted pupils also need to learn facts and learn how to do things. A knowledge base is essential for conducting any mathematical enquiry. I think that the problem arises from the amount of time a teacher employs this model of instruction. If most of the lesson is conducted in this style, it can leave a 'gifted' mathematician, who may know much of what is being taught or is able to think at higher levels, bored and frustrated. For developmental psychologists, who maintain that intellectual development is enhanced by the student actively learning, the osmosis style of teaching will leave the learner deficient in intellectual challenge and effective learning.

The teaching style which encourages *parenthesis* will involve some discussion between pupils and the teacher. It will involve some interaction between the teacher and pupils and some open-ended questioning. A certain amount of negotiation of learning will take place during the teaching sessions. Instruction will include whole-class teaching of problem-solving skills and the application of formulae. Pupils will construct their own understanding from what is said in the classroom and place it in a personalised enclosure of knowledge.

This style of teaching, again, will predominantly involve whole-class teaching. It is not as passive as the osmosis style, as some cognitive restructuring can take place. It may also inhibit pupils from following their own ideas and investigating them further. In this model, the teacher decides what to teach even in the context of problem-solving skills. Take a moment to reflect on the features of mathematically able pupils. They are able to spot patterns, pursue their own lines of thought and produce unique and elegant solutions. A style which is interactive, but controlled by the teacher, is unlikely to encourage able mathematicians to produce creative and original solutions. The parenthesis style has its place in mathematics teaching. A teacher demonstrating ideas, asking questions and encouraging explanations makes this a more active style of learning and some constructions of knowledge are possible.

The teacher who adopts the third style, which I call the *metamorphosis* style, will enable the learner to be engaged in a creative transformation of knowledge, producing conjectures and personal theory making. The learner will be given some freedom of enquiry by being asked to be engaged in investigative work individually or in groups. The teacher takes a facilitative role to enable the pupils to put forward their own hypotheses, test them and make generalisations. This style of teaching does not mean non-directed free play, which assumes the self-discovery of mathematical ideas. The teacher provides training in mathematics processes and sets challenging tasks which will direct the pupils to research for new facts and be engaged in refining their ideas. For very able mathematicians this style of teaching opens up new horizons. The time for pursuing individual enquiries can be released from some kind of compacting of the repetition of the content, mainly in textbooks. Efforts will be made, inspired by the intrinsic pleasure of creating new knowledge. Many of the models developed in the USA for mathematical enrichment have elements of metamorphosis in them.

It may be useful for a teacher to take account of the features of the tripod model. During a mathematics lesson one needs to integrate the three styles with the individual needs of the learner guiding the balance which needs to be achieved. Just as a real camera needs the tripod to retain focus the child needs the support of all three styles of teaching to enhance the brightness of the picture he or she takes of mathematics. In practical terms, in a daily mathematics lesson too much whole-class teaching can achieve only a mastery of facts and skills and some understanding. Individuals and groups must be provided with opportunities to be engaged in tasks which involve higher order skills of analysis, reasoning and evaluation. This is necessary not only for acquiring greater competence in mathematics as a discipline, but is also vital for the development of positive attitudes to the subject.

You may find it interesting to consider what has been discussed above in the context of what the Framework suggests for 'catering for pupils who are very able' in mathematics.

> Nearly all able pupils will be taught with their own class, whether it be a higher ability set or a mixed-ability class . . . They can be stretched through differentiated group work, harder problems for homework, and extra challenges – including investigations using ICT.   (DfEE 1999b: 20)

It also suggests that special arrangements may have to be made for 'exceptionally gifted' pupils, who can follow individualised programmes at appropriate times in the daily mathematics lesson, with far fewer practice examples and many more challenging problems to tackle.

Learning mathematics with understanding

In the previous section I discussed teaching styles. I argued that a teaching style which predominantly involves a passive role in which knowledge is considered to be transmitted by the teacher to the pupil is unlikely to enrich an able pupil's mathematical talents. A *constructivist* model, which has gained popularity in recent years, is based on the belief that a learner constructs his or her own learning from a variety of sources, including teacher input. The *constructive* learner personalises mathematics through reflection on his or her own learning, resolves conflicts created through self-questioning and through being asked challenging questions by the teacher. The features of able mathematicians discussed in previous chapters would suggest that a constructive style of learning would suit the more able mathematician.

Children constructing their own learning should lead them to a better understanding of mathematical concepts. A robust understanding of mathematical ideas would enable them to succeed in their pursuit of more challenging and complex ideas. An excellent exposition of the role of understanding in mathematics is provided by Skemp (1976), which makes worthwhile reading for all teachers. Only a brief summary is provided here.

Skemp describes two types of understanding in mathematics learning: *instrumental understanding* and *relational understanding*. Instrumental understanding is based on learning rules such as the rules for dividing fractions, multiplying negative numbers and finding the area of a rectangle. The results are 'immediate and more apparent'. Skemp lists the reasons why this kind of understanding may be favoured. Examination results and over-burdened syllabuses are two such reasons. The advantages of relational understanding are listed as: it is more adaptable to new tasks; it is easier to remember; it is more satisfying for the learner; and the relational schema are organic in quality. These four advantages stem from the fact that pupils who have developed relational understanding have done so by making connections with the existing framework of knowledge.

I will use one of the examples provided by Skemp to illustrate the advantage of aiming for relational understanding. Pupils who develop instrumental understanding learn and possibly remember the rules for finding the areas of different shapes, which they may see as unconnected rules. So they will learn separate rules for finding the areas of triangles, rectangles, parallelograms and trapeziums. A student with relational understanding will see all these in relation to the area of a rectangle. When the need arises, even if you forget the rule you can remember the connections and either recall the result or find alternative ways of working out the area of a shape.

Perhaps it is the emphasis on the quality of learning encouraged by a teacher who aims to develop relational understanding in his or her pupils which makes it immensely appropriate in the context of provision for able pupils. Skemp describes the benefits:

> if people get satisfaction from relational understanding, they may not only try to understand relationally new material which is put before them, but also actively seek out new material and explore new ideas, very much like a tree extending its roots or an animal exploring new territory in search of nourishment.
> (1976: 24)

What are the implications for a teacher who encourages relational understanding? What evidence would be seen in a teaching session which emphasises relational understanding? Try to make a list. Personally, I think the following features would be present:

- There will be discussion between teacher and pupils and between the pupils themselves.

- There would be some opportunities for pupils of similar ability to work together sharing and challenging ideas.
- The teacher would be involved in higher order questioning: probing questions such as: 'How do we know this?'; 'Why did you select that method?'; 'What are the advantages?'; 'Can you find the right proof from this collection of proofs?'; 'Here is a pupil's work. He has listed all the stages in his thinking. Is he right? If not, where is the mistake?'; and 'Can you tell me about some of the ideas you tried which did not work?'
- Pupils will keep mathematical journals where they record their mathematical adventures, their lines of thought, explanations and outcomes.
- Practical activities would be meaningful and would support a formalisation of ideas.
- There will be analysis of errors and misconceptions.

The National Numeracy Strategy provides an excellent framework for all the above features to happen in mathematics classrooms.

Organisational structures

Setting

What are the best ways to organise teaching which maximise the learning opportunities for mathematically promising pupils? One of the options available is *setting* pupils according to their mathematical ability. A recent government White Paper, *Excellence in Schools* (DfEE 1997) puts forward setting as a way forward for primary schools. What do we mean by setting? As literature on setting is sparse, HMI carried out a survey (OFSTED 1998) exploring aspects of setting. What do we mean by setting? OFSTED offers the following definition for their survey:

> setting is defined as the formation of teaching groups for a particular subject based on the pupil's prior attainment in that subject. Sets thus formed are therefore different from, and may be larger or smaller than, the pupil's usual classes. Sets may contain pupils from more than one year group. Setting is different, therefore, from the formation of ability groups, taught by the same teacher, within mixed ability classes. Setting is also different from streaming, where pupils of similar ability are taught together for all subjects. Streaming is rarely found in primary schools today.
> (1998: 1)

In practice, schools set children in groups which are often described as 'top', 'middle' and 'lower' sets. Some schools with three parallel classes may set children into three groups; others may create extra groups. This choice usually depends on the availability of extra staff.

How does setting help able pupils?

The main reason for setting is to reduce the complexity of teaching groups of children with a wide range of ability. Teachers often experience difficulties in keeping a balance between supporting pupils who are struggling, and at the same time, wanting to provide opportunities for able pupils to learn more advanced concepts or be involved in in-depth exploration of ideas. In many cases, they opt to teach with the 'average' child in mind. In a top set, the ability range is narrower, which makes differentiation easier to manage, where teachers can select tasks and resources which can be tackled by most pupils.

In a 'setted' class, pupils of similar ability can discuss ideas together much more easily. It is easier to provide challenging activities and discussions in this situation. Research (Kulik and Kulik 1992) shows that grouping by ability helps higher ability learners. The pace of instruction can be increased without having to consider whether the whole class has understood what has been taught. It is also easier to ask both closed questions, which test knowledge, and questions which involve higher levels of thinking to a set of pupils who have higher capabilities for learning and for reflecting on their learning.

One of the problems often articulated by teachers who attend our in-service courses is their perceived inadequacy with their mathematics subject knowledge. In many schools this problem is resolved by getting the mathematics co-ordinator, or someone who enjoys teaching mathematics and feels she has the subject expertise, to teach the top set.

Teachers often say that they find it easier to assess children's learning, their special interests, learning styles and strengths within 'set' groups. Teaching pupils in 'sets' also enables teachers to achieve a greater degree of match between learning tasks and pupils' aptitudes. It may be easier to conduct productive plenaries with pupils within a 'set', which may have progressed sufficiently with ideas; plenaries can be difficult to manage in mixed ability groups. Managing children's individual interests is also easier within sets.

It is wise to remember that setting alone does not guarantee 'good practice' in teaching the able. Setting needs to be organised carefully. The following may be useful points for consideration:

- Use a range of assessment information when selecting pupils for the top set. National Curriculum and other test scores should be supplemented by consideration of qualitative attributes. It is likely that an exceptional mathematician may not do well in tests for all sorts of reasons, not least of which is that he or she is bored with the tests. When I visited a test centre for gifted pupils in the USA, I was told that the most able mathematicians did not perform particularly well in multiple choice tests as they did not feel comfortable with the 'closed' nature of the solutions and sometimes added other options as choices. As the tests were marked by computers their original and creative answers were marked wrong! In Chapter 2, I introduced a variety of sources which can be drawn on when assessing children's abilities.

- Have flexibility within groups. Sometimes it is difficult to identify pupils with high mathematical ability within a short space of time. Late developers and those who may hide their mathematical talent for fear of being teased or bullied should be moved to higher sets when their talents have been spotted.
- Mathematically more able pupils in the top set should be given activities which match their aptitudes. Giving them more of what the other groups are doing or using a textbook designed for a higher age group will not extend their mathematical thinking. The features of 'good' activities – presented in Chapter 4 – will provide some guidance.
- One of the drawbacks of setting is that the rest of the class misses out on their expertise; this can be addressed by arranging occasions when the top set can make presentations of their ideas or be responsible for organising interactive displays or designing mathematical games and quizzes for the rest of the class and, of course, for younger children.

In some cases, variations in setting may be used as suggested in the guidance (DfEE 2000b) offered by the National Numeracy Strategy. They are:

- temporary setting, for example during revision sessions;
- part-time setting, say two or three times a week;
- have a top set, but teach the rest of the children in mixed ability groups;
- set the children as they get older.

But caution should be exercised. We are told that if setting is across two years, we need to ensure that work is not a repeat of the previous year.

Other organisational options for providing for able mathematicians

As pupils vary in their abilities, interests, learning styles and emotional stages, it is not possible to recommend one specific method for providing for all mathematically promising students. As schools have different organisational structures and different philosophies and practices, I will briefly discuss the main options available for consideration and the making of decisions. These options have been in practice in other countries, mainly in the USA, for many years. I will draw on the US evidence and my own experience to provide an overview of the different practices.

The Centre for Talented Youth (CTY) in Johns Hopkins University has been engaged in research and development in issues relating to the identification and nurturing of mathematical talent since 1979. The 'optimal match' philosophy of the centre (CTY 1994) is a worthwhile starting point for a discussion of the options available to us.

> An 'optimal match' is the adjustment of an appropriately challenging curriculum to match a student's demonstrated pace and level of learning. The 'optimal match' must be preceded by an evaluation of a student's interest and ability in a subject. For example, let us take a third grade student with a demonstrated ability in mathematics beyond third grade. This student is bored, frustrated and perhaps exhibiting behavioural problems. Once the degree of ability is ascertained, a variety of educational strategies are available to adjust instruction properly to the child's situation. Among these options are placement in a higher grade for mathematics, relevant enrichment in the third grade for mathematics, an out-of-classroom mentor in mathematics with the child pursuing other relevant activities during the math period and a summer school program.

It goes on to say that the option chosen in the optimal match process depends on a number of factors including an accurate assessment of the educational needs of the child, a consideration of parental support and reinforcement and the resources available in the local community.

The three options available – acceleration, enrichment and working with a mentor – are now discussed in turn. It should, however, be borne in mind that these options are not mutually exclusive. For many able children combinations of these may offer the most appropriate action.

Acceleration

Acceleration means learning the same content at a faster pace than other pupils in the same class or peer group. For example, a 9-year-old may be given opportunities to learn mathematics from a higher level in the National Curriculum, say from Level 6 or 7. This is sometimes organised by allowing pupils to learn mathematics subjects with older pupils, or by moving them up one, two or three years or by providing them with a special programme within the classroom which includes more advanced ideas than those expected to be learnt by other pupils. In some cases a more radical acceleration is adopted. For example, children aged 8 or 9 preparing to take GCSE examination in mathematics. This kind of radical acceleration is, however, relatively rare in the UK. Those who recommend this method of provision believe it to be a viable option for pupils who are fast learners. It avoids a pupil being bored with work he or she has already mastered.

Research carried out at Johns Hopkins University and by Gross (1999) in Australia over two decades, indicates that acceleration may be the most suitable option for some pupils. Gross recommends that schools should not refuse to use this method of provision without careful consideration. She cites examples of many pupils who felt challenged and fulfilled when they were moved up a year or two.

Those who advocate acceleration argue in favour of this strategy as they maintain that it enables fast learners to master ideas at a faster rate by using teaching material which matches their ability. Being 'moved up' a few years is seen to be a recognition of their capacity to learn fast. This may sometimes boost pupils' motivation and improve their behaviour and enhance achievement.

It is hard to list any possible benefits of acceleration as there are no recent research studies in the UK to provide any evidence. In his illuminating discussion on the topic of acceleration, Fielker (1997) maintains that through acceleration pupils 'do not learn more about mathematics. What they do is merely learn the same mathematics sooner'. This he says 'does not seem to fulfil the needs of the more able, who deserves something better'.

Fielker lists the possible problems of schools adopting acceleration as a strategy. The problem of maturity is of some concern. Younger children may be brighter, but they may have problems mixing with children who are 'possibly physically, socially and emotionally mature'. Then he mentions the possible danger of resentment felt by ordinary and less able pupils in a class where a younger person is appearing to do better. This could be greater than the resentment felt by pupils about brighter pupils in their own class.

If a school is considering acceleration as a strategy, it should be on the basis of evidence gathered by the teacher through observation of the pupils and extensive consultation with parents. At the same time, it is wise not to rule out this strategy completely without considering its possible advantages for certain pupils if they are moved up a year or so.

During my discussions with teachers, it often emerges that teachers who adopt this strategy do so in the absence of something better. The question of 'what next for the pupil who has done GCSE early' is also frequently asked. Teachers frequently refer to the absence of in-service programmes targeting the topic of provision for mathematically able pupils. Lack of appropriate resources is also cited as a reason for adopting acceleration as a strategy.

Freeman's (1998) discussion on issues of acceleration will provide useful reading for those who wish to consider aspects of acceleration in more detail.

Enrichment

An alternative strategy to acceleration is to provide opportunities for pupils to learn mathematics topics in more depth and hence go beyond the acquisition of facts and skills. It involves a broadening of the knowledge base as well as the processes of learning.

Provision through enrichment activities has many supporters. While being engaged in enrichment activities, pupils will have opportunities for widening their knowledge base, which is one of the aims of using the acceleration strategy. Sheffield's (1999) discussion on the frequently heard analysis of the mathematics curriculum in the USA that 'it is an inch deep and a mile wide', is worth taking note of, also, within the UK context. Sheffield goes on to say:

> Mathematics textbooks in the United States tend to cover large numbers of topics at a relatively shallow level and repeat the same topics for years . . . This is especially detrimental to good mathematics students who have already mastered content of the mathematics program and are bored with the repetition.
> (1999: 16)

Whether to accelerate or enrich students in mathematics lessons continues to be a topic of debate in the USA and, more recently, among UK educators. A three-dimensional model is suggested by Sheffield for adoption when considering effective provision for able mathematicians. She describes the model (Figure 5.3) as an attempt to illustrate that services for our most promising students should not only look at the rate of presentation or the number of mathematical topics, but must also look at changing the depth or complexities of the mathematical investigations.

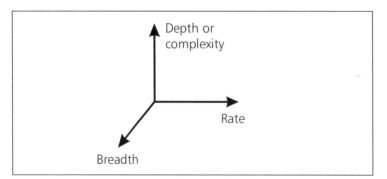

Figure 5.3 The three-dimensional model of mathematical learning (Sheffield 1999)

What mathematically able pupils need, Sheffield maintains, is time and encouragement 'to explore the depth and complexities of problems, their patterns, and connections among them'. This kind of programme is not about introducing a puzzle a week or the occasional extension task; it must become part of the regular mathematical diet.

Lack of resources and exemplar models of enrichment activities are often cited as reasons for not adopting enrichment strategies. The pressure of having to teach the National Curriculum and the structure of the National Numeracy Strategy are also used as reasons for not setting up enrichment activities. In fact, what they both provide are structure and content; it is up to the teacher to offer enriching opportunities using what these documents contain as starting points.

Working with a mentor

If you have pupils who demonstrate exceptional ability in mathematics you may consider finding a mentor; an adult who can provide support. Whether this support is provided in the classroom or outside the classroom would depend on the extent of the pupil's ability. A mentor is likely to have a strong interest and expertise in mathematics. The mentor is a resource person who may be selected from a local secondary school, university or from the local community. A 'good' mentor will enhance the pupil's intellectual development and should have the following desirable characteristics:

- be older than the pupil or be an adult;
- be interested in aspects of learning of the pupil;
- motivate the pupil;
- be very enthusiastic about the subject and act as a role model;
- be able to support the pupil with subject expertise.

There is theoretical support for the role of an adult in exploring the true potential of a child. Vygotsky defines his zone of proximal development as:

> the distance between the actual developmental level as determined by independent problem solving and the level of potential development as determined through problem solving under adult guidance.   (1978: 86)

Vygotsky believes that a child's state of mental development can be determined only by clarifying its two levels: the actual developmental level and the zone of proximal development. In that context, a mentor has an important role to play in both identifying the child's potential and supporting him to maximise future mathematical achievement.

Differentiation strategies

Effective differentiation is the key to raising achievement. Differentiation can be achieved through tasks, through providing opportunities for higher levels of thinking and by selecting appropriate teaching styles and resources. Many of these aspects have been addressed in previous sections of this book. Both experienced teachers and trainee teachers find differentiation a real challenge. When I study lesson plans, I notice that many teachers interpret differentiation as catering for children who have learning difficulties, but we also need to consider how we can meet the needs of our most able mathematicians.

To start thinking about differentiation, think of a pupil in your class who finishes the work faster than others. You may want to do this with a colleague. Reflect on the most recent lesson you taught when that pupil 'finished' the set work. What strategy did you use to extend that pupil's learning? Ask yourself whether you are happy with that strategy?

Now, make this exercise even more interesting and useful by examining the following list of strategies observed in classrooms. Again, you may want to do this with a colleague or a friend or in a year group, say when you are planning children's work for the week. Tick the responses you feel will help to extend the child's ability and put a cross next to the ones you think are inappropriate.

> A Year 3 child says, 'I have finished this work, what shall I do now?'

- 'See if you can find some more pages on the same topic. Can you do them?'
- 'Go away, I am busy right now.'
- 'Find some children who are struggling and help them.'
- 'How did you get on with that investigation? Bring me your diary so we can think of some new ideas for you to work on.'
- 'As long as you are quiet, you can do something else.'
- 'Look in our challenge box and choose something you like doing.'
- 'Do some reading now.'
- 'Can you think of a different way to tackle that problem?'
- 'What if you were telling someone else how to do it. Write down what advice you would give that person.'
- 'Can you design a new activity based on the one you have just done?'
- 'Tidy up the shelf; it's a mess.'
- 'I don't know what to do with you. I will borrow the book Year 4 are using.'

The purpose of this activity is to highlight the need to be prepared for the child beforehand, and not have to think on your feet. I will conclude this section with some practical strategies for differentiation. Some of these strategies achieve differentiation by task, others by outcome.

- All children need to learn facts and skills and be taught mathematical ideas such as place value. These can mostly be achieved by teaching the whole class together using differentiated questioning strategies.
- During the week, select some tasks which have open-ended outcomes and ask pupils to explore ideas at an appropriate level matching their ability. If pupils are reluctant to try harder work, you may need to offer extra rewards for undertaking challenging activities.
- Make problem solving and investigations a regular feature of mathematics lessons; multi-level enquiries are easier to organise during these lessons.
- Have puzzles and other challenges available in the classroom.
- Don't discuss the conclusions of all the activities during the plenary on the same day. Ask children to think of a question which they are wondering about in the investigation they are carrying out. Extend the time for a particular piece of work or ask pupils to do it as homework.
- Give opportunities for checking proofs, calculations and generalisations.
- Let children follow an individual assignment, if the rest of the class cannot manage it.
- Prepare a group for taking the world-class test in mathematics and/or problem solving.
- Use ICT for extending ideas and for presentation formats.
- Use distance learning packs for individual pupils.

Out-of-school opportunities

Increasingly, these are being made available to pupils. Summer schools, university based extension programmes, contests, after-school clubs and correspondence courses offer opportunities for pupils who have the ability and the stamina to study mathematics in more depth. Access to the Internet also adds to the number of opportunities. Chapter 7 will include more details on resources for the mathematically promising.

A Square Deal for able pupils

To conclude this chapter I will draw on the CK model in Figure 5.4, which explains many of the aspects of provision described in this chapter.

Although the diagram is self-explanatory, I will explain the thinking behind the model. All children have an entitlement to learning the statutory content of the mathematics National Curriculum. Within the classroom, the offered curriculum must be enriching in experience and be differentiated to match the ability of all children. Within this organisation it may be necessary to make adaptations, say setting, to make the ability range narrower. Within the top set children need to be provided with opportunities to explore mathematics in depth, not more of the same. Special provision needs to be considered for children who demonstrate exceptional talent in mathematics. Acceleration is

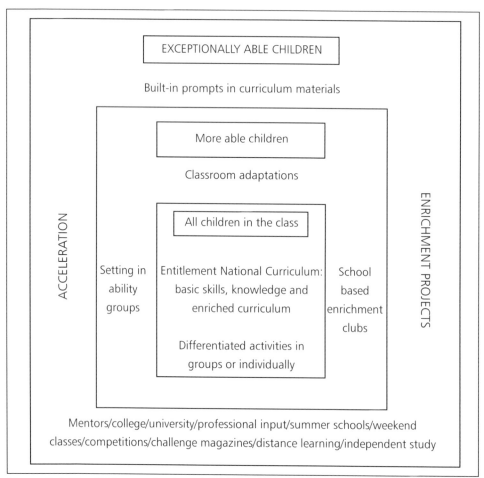

Figure 5.4 The CK Square Deal model for provision for able pupils (Koshy and Casey 1997)

one such strategy, extended projects within school hours or outside school should also be available. They should also be considered for extra support, as shown in the outer square, through mentors, summer schools, university based schemes and so on.

Summary

In this chapter I have considered aspects of organisation in the context of educating able mathematicians. First, four critical factors which affect student learning were listed. They are: assessment, teaching styles, organisational approaches and differentiation. The important issues relating to each of the above were discussed and practical strategies were explored. The most important message I wish to convey to the readers is that it is unlikely that any single strategy or option will meet all the needs of mathematically able pupils. Being aware of the options and suggestions introduced in this chapter should support the teacher to make informed choices to help mathematically able children to fulfil their potential.

CHAPTER 6

Using ICT to teach able pupils

In this chapter we will explore ways in which ICT can be used to teach able pupils. What do I mean by ICT? Calculators, computers, video recorders, tape recorders and all multimedia resources can be listed as ICT. All these can be used in the teaching of mathematics. The next question is why focus on ICT with reference to able pupils? Take a moment to scribble down some reasons on the basis of what you know about the special attributes and needs of the most able mathematicians, as discussed extensively in the preceding chapters. How do your reasons compare with the following list, which was produced by a group of teachers on an in-service programme?

- ICT offers a valuable tool for doing mathematics. The power and speed offered by computers make them suitable for use with able mathematicians, who can process information quickly.
- Computers and calculators are very useful for able pupils who may be working on tasks independently.
- ICT is useful when pupils are engaged in problem-solving activities. They can carry out enquiries using real or purely mathematical contexts, letting the machines do routine calculations. The time and mental energy saved can be utilised to develop problem-solving processes.
- Able pupils often have a flair for identifying patterns, making conjectures and generalisations. Computers and calculators enable them to test their conjectures efficiently and to generate formulas.
- Able mathematicians will find both pleasure and challenge in using software which can be used to experiment with shapes and create geometric patterns.
- There are many websites on the Internet offering challenging mathematical activities and details of national and international competitions. These offer extra challenges and enrichment opportunities for able pupils.

- Both the National Curriculum and the National Numeracy Frai
  require pupils to be proficient in the use of ICT.
- Able pupils are often asked to take a leadership role in designing pi
  organising displays and quizzes. The presentation of these ca
  enhanced by using word-processing and art packages.

In the following sections I will present some practical examples of how
teachers have set challenging activities for mathematically able pupils using
calculators and computers. Although these are discussed in the context of
teaching able pupils, these motivating and challenging activities can be used
with all pupils at different levels of difficulty.

Working with a calculator

There has been much debate, in the past 20 years, on the use of calculators in
mathematics lessons. Many of us are familiar with articulated views ranging
from 'calculators cause brain damage' to 'calculators enhance pupils' mathematical understanding'. I have discussed various viewpoints on the use of
calculators elsewhere (Koshy 1999), but for the purpose of this chapter I have
selected examples in which a calculator is used to encourage pupils to:

- investigate number patterns and sequences;
- understand principles underlying the number system;
- look deeper into mathematical ideas, rules and procedures;
- use estimation and reasoning skills;
- extend existing mathematical knowledge;
- discuss and share ideas.

Using calculators to investigate number patterns, sequences and number properties

One of the uses of the calculators emphasised in the National Numeracy
Framework is to explore number patterns and number properties.

By using the 'constant' function children can find doubles and halves and
predict number patterns and sequences. Put in number 2, then press ×
followed by 4. The display shows 8. Now each time you press =, the calculator doubles the number and gives you, 16, 32, 64 and so on. Try 248 divided
by 2 and press = to see the reverse happening. Let children use this function

to generate number patterns and sequences at their own level. Using the idea of 'function', pairs of children can set challenges for each other.

Divisibility rules

A class of Year 4 children used a calculator to investigate divisibility rules. The teacher asked the children to investigate whether the following rules of divisibility always worked. She realised that this involved working with larger numbers and doing many calculations.

A number is:

- divisible by 2 if the last digit is even;
- divisible by 3 if the sum of the digits is divisible by 3;
- divisible by 4 if the total of the last two digits is divisible by 4;
- divisible by 5 if the last digit is a 5 or 0;
- divisible by 6 if the number is even and is divisible by 3;
- divisible by 9 if the sum of the digits is divisible by 9;
- divisible by 8 if the sum of the last three digits is divisible by 8.

As the purpose of the lesson was to explore the ideas of divisibility, the calculator helped to release children from some of the tedious calculations which would have stood in the way of free and enjoyable enquiry.

Calendar patterns

A group of 6-year-olds were investigating number patterns using calendars.

Look at the page of a calendar. Choose a 2 by 2 square and add the number in the bottom left corner to the number in the top right corner. Now try adding the two numbers on the other diagonal as shown. See what happens.

- Try doing this with other sets.
- Why do you think this is happening?
- Try a 3 by 3 square
- Think of some other patterns you can discover.

Digit sums

A group of Year 4 pupils investigated the following.

If you add two numbers, will the digit sum of the answer be the same as the digit sum of the digit sums of the numbers.

$$
\begin{array}{ll}
567 & \text{digit sum} = 5 + 6 + 7 = 18 = 1 + 8 = 9 \\
+\ 314 & \text{digit sum} = 3 + 1 + 4 = 8 \\
\hline
881 & \text{digit sum} = 8 + 8 + 1 = 17 = 1 + 7 = 8
\end{array}
$$

Add $9 + 8 = 17 = 1 + 7 = 8$

The digit sum of $881 = 8 + 8 + 1 = 17 = 1 + 7 = 8$ also

- Does this always work?
- Try it with other operations

How would you use this method to check your calculations?

Happy numbers

This investigation was carried out as a whole class activity by a Year 4 class using calculators. The more able pupils were further challenged to find all the happy numbers from 1 to 100.

Select a number, say 19. Square each of the digits and add their squares.

$$1^2 + 9^2 = 1 + 81 = 82$$

Square the digits of 82 and add them together.

$$8^2 + 2^2 = 64 + 4 = 68$$

Square the digits again and add them together.

$$6^2 + 8^2 = 36 + 64 = 100$$

Square the digits and add them together.

$$1^2 + 0^2 + 0^2 = 1$$

It ends with a 1. So 19 is a happy number.

Find some more happy numbers. What do you notice about the sad numbers?

Developing a number sense

In the following activities the role of calculators is not primarily to perform calculations, but to help in encouraging children to think, in depth, about what is involved in calculation.

Six discrimination
Yvonne used the 'Six discrimination' activity (Figure 6.1; NCC 1992) to encourage a group of Year 5 pupils to think about number operations and to develop reasoning, decision making and communicating skills. Pupils were asked to share a calculator and to work in pairs and agree their method before they recorded the results of using the method. This specific strategy was introduced to encourage pupils to share their thinking.

The activity which generated a great deal of mathematical thinking and 'number talk' can be adapted for all age groups.

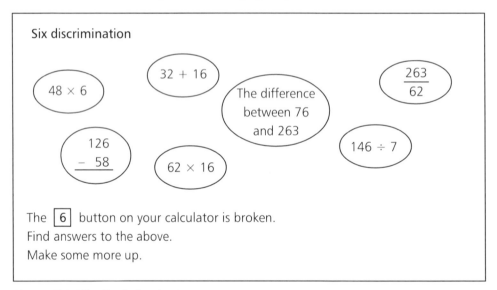

Figure 6.1 'Six discrimination'

Broken keys
As in 'Six discrimination', here again, the role of the calculator is to facilitate thinking about aspects of number operations, estimate and reason.

Using only the keys 4, $-$, $=$ and $\times$, can you make all the numbers 1 to 20?

Can you make 100 using only 3, 7, $+$, $-$ and $=$?

Using only 4 and any operations can you make all numbers up to 10?

As a stimulus to extending knowledge

Calculator mystery

Mystery 3751(a SMILE activity – see Resources in Chapter 7) involved using a calculator to work out calculations and then to use the upside readings of the display as clues to solve the mystery (Figure 6.2). This activity was used by a Year 7 class as a stimulus to encourage pupils to revise and reinforce procedures for calculations.

Pupils were highly motivated by this activity and when the teacher, Diane, extended the activity it also provided a real challenge to the most able pupils. She asked her pupils to construct a similar map to find a 'missing something', for example, a 'treasure map'. Further, she asked them to include at least one calculation involving long multiplication, percentages, fractions and decimals in their clues.

Find the largest product

Using the digits 1, 2, 3 , 4 and 5 find the largest product using the format;

Try the following format

☐ ☐ ☐ ☐
×   ☐

This activity usefully kept a group of Year 5 able pupils busy for 20 minutes or more.

Solving 'real' problems

Daniel has been alive for 70128 hours. How old is he?

A small packet of biscuits costs 85p and it weighs 930 grams. A larger size packet costs £2.79 and it weighs 3.1 kilograms. Which is a good buy? Explain your reasons.

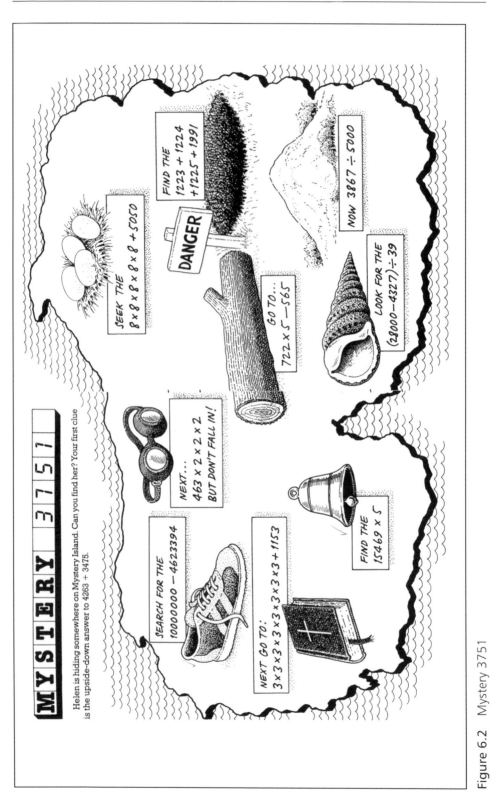

Figure 6.2   Mystery 3751

In the previous contexts, a calculator helps children to tackle problems which involve working with large numbers and complex calculations. The use of a calculator enables a pupil to place the emphasis on methods for solving the problem and not be put off by the need to do time-consuming calculations.

Other interesting and challenging data-handling problems, such as finding the percentage distribution of vowels on a newspaper page or the average number of cornflakes in a packet, become more accessible if the pressure of tedious calculations is taken away from the situation.

Problems which involve developing an understanding of number

The following collection of problems were part of a 'mathematics challenge box' in a Year 6 class. Calculators were allowed for solving them. Try them and reflect on the processes involved. You should see why they are classified as 'challenges'.

Three consecutive numbers multiplied together give 6,635,970. Find the numbers. How many tries did you have?

Now try 778,596, which is also the product of three consecutive numbers. How many tries did it take you to find the three numbers?

I put a number into my calculator and multiplied it by another number. The answer was 377. What were my numbers?

If you found that too easy, try this:

I divided one two-digit number by another two-digit number and got 0.78125 as the answer. What were my numbers?

Calculator games

An abundance of calculator games is available. Many of these are designed to practise mathematical operations. It is likely that a mathematical game you have selected for most of the class may not be challenging enough for the most able pupils in the class. For example, a 'Blockbuster' type of game designed to practise the multiplication of whole numbers may be too easy for one group. In this case you may want to differentiate by giving them a more sophisticated version of the game. I found the following version of the game (Figure 6.3) in a mental mathematics resource pack (Koshy 1999) useful in providing a

Decimal connect

A game for 2–3 players

You need: different-coloured counters – one colour for each player, a copy of the grid below

- Take turns to select a number on the grid.
- Choose two numbers, one from the circle and one from the rectangle and multiply them to give a number on the grid.
- If the calculation is accepted as correct by the other players, cover the grid number with a counter.
  (In case of disagreement use a calculator to check the answer.)

The aim is to get four counters in a line, horizontally, vertically or diagonally to score a point. Keep playing until all possible spaces are covered.

0·3	25	21·09	0·84	300	2
14·06	60	1·92	20	6	0·04
1	35·15	2·4	0·1	500	12
0·6	3	20	50	800	4·8
1·68	8·4	0·8	2·8	4·921	1·2
70	70·3	0·72	49·21	1	56·24

Circle: 5, 3, 2, 0·5, 0·2, 10, 0·7, 8, 0·25

Rectangle:
12	1·5
1·2	0·2
100	0·25
2·4	4
7·03	0·5

Figure 6.3 'Decimal connect' (Koshy 1999)

motivating and challenging activity for a small group of Year 4 children when the rest of the class was playing a version which involved multiplying two-digit whole numbers by a single digit whole number.

A further way of challenging able children is to ask them to design calculator games for the rest of the class which would involve them in designing the game, organising the ideas and checking the solutions. It is far more challenging than just playing a game themselves.

In the previous section I have tried to demonstrate the potential of using calculators to extend children's thinking. As mathematically promising pupils are unlikely to need hours of practice and reinforcement, the time released can be utilised for exploring challenging mathematical ideas using a calculator. We have considered many of the advantages in using calculators as resources with able pupils. One further advantage of using a calculator is that pupils can get on with their explorations without a teacher having to check their answers all the time.

Finally, I saw a very interesting project on 'Calculators' in a Year 7 classroom. A group of able pupils studied different types of calculator available and wrote leaflets on 'how to use' them giving examples of activities. Another group was in the process of writing a book entitled 'Calculator puzzles you are sure to get addicted to'!

Using computers to extend able mathematicians

In this section we will explore the potential for using computers to extend and enrich pupils' mathematical experiences. Since the introduction of the Framework, the way teachers use computers has changed. The Framework offers the following useful advice on the use of computers, which is particularly relevant in the context of provision for able pupils.

> You should use computer software in your daily mathematics lesson only if it is the most efficient and effective way to meet your lesson's objectives. For example, an aimless exploration of an 'adventure game' or repetitive practice of number bonds already mastered, is not a good use of lesson time.
>
> (DfEE 1999b: 32)

A good use of the computer is to extend pupils' mathematical thinking and understanding. It can also be used to make tasks easier and to create mathematics.

When we consider computer software it is useful to think of the types available to us: programming languages such as Basic and Logo; structured programs which are designed to introduce, practise and consolidate mathematical ideas; and content-free software, which is a useful tool for conducting surveys by analysing and interpreting data. Problem-solving activities are also available.

In the following pages, I will consider the ways in which a computer can be used in order to maximise the learning opportunities for mathematically able pupils. The focus is on the ideas, not on the technical side of computer usage. It is assumed that teachers who intend to use these ideas will familiarise themselves with the software packages.

Using programming languages

Using Basic, pupils can write programmes to generate number patterns and test and prove generalisations. They can also create *fractals*, which are challenging and visually pleasing.

Logo is the most frequently used and is the easiest of the programming languages for primary school children.

Seymour Papert (1980) described Logo as a tool which enables pupils to think. The benefits of using Logo with pupils are:

- Pupils can write programs to produce geometric patterns which can be visually pleasing and satisfying. When children are more advanced they can use Logo to write programs to generate number patterns and design adventure games, mazes and puzzles, all of which are very challenging and are particularly suitable tasks for able mathematicians.
- Writing programs using Logo requires pupils to communicate mathematics to the turtle; this involves thinking about mathematical ideas in depth and being 'real' mathematicians. Papert puts this in context very clearly when he says that when this kind of communication occurs, children learn mathematics as a living language. He goes on to say that the idea of 'talking mathematics' to a computer can be generalised to a view of learning mathematics in 'Mathland'; that is to say, in a context which is to learning mathematics what living in France is to learning French. In the context of providing for able mathematicians, these words are particularly significant.
- Logo helps children to experience useful problem-solving strategies. In order to carry out a task they have to think logically, break the tasks into manageable parts, refine strategies and check and refine ideas. As discussed in preceding chapters, all these processes help to make children more effective problem solvers. Another useful strategy acquired during Logo programming involves the need to accept mistakes as part of improving one's learning and producing more effective outcomes. Pupils learn to 'debug' their programs in order to create better and more elegant solutions; Krutestskii lists these as attributes which able mathematicians are capable of.
- Logo gives pupils opportunities for carrying out open-ended investigations. They can extend tasks by posing their own questions and setting

their own goals. This is particularly relevant for able pupils who are often capable of working independently.

I have listed only some aspects of Logo which make it a suitable resource for providing pupils with mathematical challenges. In the following sections I will describe just a few activities that I have used with different age groups. They are only samples, but as the potential for using Logo is high, these activities may trigger off more ideas. I have tried to present the ideas with some progression in mind, but in a book dealing with provision for able pupils the kind of progression I have in mind can only be arbitrary and certainly is not age-related. You may refer to the ideas in *Making IT Work for You* (Koshy and Dodds 1995).

Working with floor turtles
Playing with floor turtles is great fun and if the activities are structured properly they can also be educationally very challenging.

The activities may be cross-curricular to set the tasks in meaningful contexts. Try these:

- Can you get the turtle to move around a path which has obstacles? Think about the movements before trying them out.
- Can the turtle write your name? Write a program for this. Attach a pen and test it out.
- Dress your turtle in a postman's costume and let it deliver letters to a row of houses.
- Can you make the turtle do a presentation by acting a scene from a fairy tale?

Working with the screen turtle
In order to get children used to Logo commands try:

- placing an acetate sheet of a maze on the screen and ask children to get the turtle 'home', by giving instructions on the distances and angles and using Forward (FD), Backward (BK), Right (RT) and Left (LT);
- sticking a sweet on the maze and directing the turtle to the sweet using the FD, BK, RT and LT commands.

Moving on

- Here are three lines. How many different shapes can you make using them?

- Can you draw a face like this one?
  Make some funny faces.
  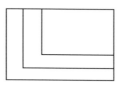

- Write a procedure to draw a square.
  Use that procedure to draw a nest of
  squares or rectangles?

- Can you draw this rocket?
  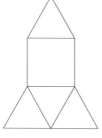

- Write a procedure to make a polygon and rotate it to make a pattern.

- Draw a five-pointed star.

- Choose a picture from a magazine and write a program to draw it on the screen.

- Can you write a program to produce
  a spiral on the screen?

- Here is a picture of an octagon, programmed by another pupil. Something has gone wrong. Can you debug it?

- Find out how you could use 'recursion' to generate some number patterns.

- Does your Logo program set coordinates? If it does, create a picture using coordinates.

Using spreadsheets

Using a calculator or a computer will save time by taking over tedious calculations and release time to be engaged in further thinking. The use of a spreadsheet to do aspects of mathematics demonstrates this very well. When you use a spreadsheet, you are doing mathematics, and not just using a computer software. A good calculator can help with calculations, but the potential support from a spreadsheet is much greater. In this section I will briefly explore ways in which the use of a spreadsheet can help pupils to be involved in asking questions, analysing, evaluating, testing hypotheses and generalising. All these processes, of course, are those we encourage in all pupils. Mathematically able pupils are often described as being more capable of applying these processes than their peers.

A spreadsheet is like a sheet of paper which can be used to calculate different things. It is set out in rows and columns called 'cells'. The software enables us to put numbers or other information into these cells and carry out investigations using that information.

I would assume that most schools have access to spreadsheets. It is useful for pupils to be shown how to use a spreadsheet, but recently I was in a classroom where the teacher asked a bright group of children to 'play with the spreadsheet' to figure out how it works. This task was certainly challenging and I was very impressed by the discoveries made by the children who worked out most of what they needed to know about using a spreadsheet. It is likely that they also learnt much problem solving from that task alone. Although there is much to be recommended in this open-ended approach to learning about spreadsheets, I think it is useful to show children how formulas and functions work. For example, as part of a survey, a formula may be used to give the total in Cell A4 when the number in cell A1 is to be added to the number in cell B2.

Here are some challenging activities you may pursue:

- Let the children work in pairs or groups. One person puts in a formula and the others work with the formula by studying the way numbers change according to the input formula. Similarly, program a function into the spreadsheet and let the others guess what the 'secret' function is.
- Use a spreadsheet to generate times tables – say the tables of 13, 19 or 28 – which fascinate most pupils.
- Produce a number pattern by typing 1 in cell A2 and then putting a formula 'A2 + 3' into cell B2 and watching what happens. Try other patterns. I have seen some sophisticated work done by pupils using this idea.
- Spreadsheets are useful in investigating hypotheses. For example, the correlation between height and shoe size or height and headband length or height and arm-span can be established with the help of a spreadsheet. The facility to produce graphs – bar graphs, pie charts and scattergraphs – adds to the usefulness of analysing, interpreting and presenting data. I have seen spreadsheets being used by pupils for carrying out surveys of all kinds, some of these during mathematics clubs. One recent example I saw was of Year 7 pupils who were investigating 'best buys' of cereals from different supermarkets. In a summer school, some Year 5 pupils were using a spreadsheet to carry out a cross-curricular task of designing and marketing an item of food. The spreadsheet helped them to carry out a mini market research survey, forecast profits and so on.

Spreadsheets help pupils to use algebra in a meaningful way. In the following example, Sutherland (1995) convincingly describes the way children carried out an algebra story problem using a spreadsheet. This kind of work will challenge the mathematically able pupil.

Chocolate problem

100 chocolates were distributed to three groups of children. The second group received four times as many chocolates as the first group. The third group received ten chocolates more than the second group. How many chocolates did the first, second and the third group receive?

The solution shown in Figure 6.4 illustrates how one of the children tackled the problem using a spreadsheet and how it helped her to think about the problem.

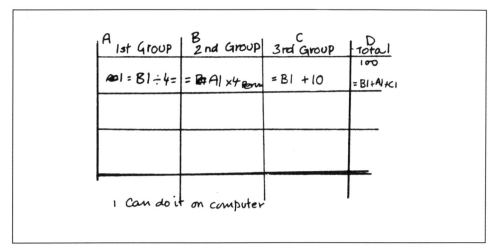

Figure 6.4 Solution to the 'Chocolate problem'

In this section I have described the type of challenges which can be set for children with the aid of spreadsheets. The intention has been to demonstrate the potential of their use rather than present an exhaustive set of ideas.

Using databases

Databases are used for storing information. They are similar to spreadsheets. While spreadsheets are more useful for handling numerical information, databases can be used to store both numerical and word information. As in the case of spreadsheets, databases also help children to sort, analyse, interpret and present data.

As far as able pupils are concerned, the power of a database makes it a suitable resource for conducting investigations, many of which may be set in real life contexts. Once children are used to setting up databases, they can use them for all sorts of enquiries. Here are some examples of using databases that I have seen in schools.

- A group of 5–6-year-olds investigated the 'favourite sweet' of the class using a simple database. This activity was carried out by a group of able children who had completed the number work they had been set by their teacher. It was common practice for Anna, their teacher, to set them work which involved a certain amount of exploration. The 'favourite sweet' was one such ongoing special project. With some initial guidance from their teacher, the pupils set up the survey of sweets, printed out graphs, put up a display and made a presentation to the class. They used graphs to show all the information clearly.

- A group of Year 5 pupils set up a database of library books in the school as part of their mathematics club work on 'real-life problem-solving'. They set up fields to input information on the type of book, author, number of pages and so on. The database helped them to sort and analyse information on the basis of which the head teacher was informed of the distribution of the type of books in the library. Subsequently, they decided to arrange a survey to find out the reading preferences of the children in the school so that they could see any mismatch between what was provided by the school and the reading preferences of the children.

- While the previous two activities were carried out as special projects, a class of 9-year-olds was involved in a whole-class activity; 'setting up a shop' as part of their data-handling lessons. All the children were involved in designing a questionnaire to do some research into what the shop should sell. Analysis of the results and the presentation of findings were carried out by the 'top group' in their mixed ability class. The presentation and displays were further enhanced by the use of a range of charts and tables.

- A highly motivating project with the title 'Which crisp?' involved a whole-class data-handling project carried out by Year 6 pupils. Again, the differentiation was achieved by assigning more complex tasks to the more able pupils. The project was launched by one of the parents, who worked for a supermarket, giving a talk. During the introductory talk, the speaker discussed the role of consumer surveys and the basic principles of setting up such a survey. This 'real-life' touch provided greater motivation for the children, who subsequently worked hard to conduct an exciting survey and disseminate their results. They decided to set up a list of criteria to be used to judge the quality of the crisps. A data sheet was designed which included headings such as feel, cost, shape, flavour, texture, taste and so on. Crisps tasting (with their brand name hidden by removing it from the packet to stop them being prejudiced in their assessment) was followed by setting up the database, which enabled them to ask appropriate questions to help decide which brand of crisps was better value for money. A display and a presentation of ideas made the project an enrichment activity for the whole school.

A careful consideration of the activities I have described in this chapter will show how the use of ICT can enhance children's learning by allowing them to develop the following features:

- asking their own questions;
- being systematic;

- sorting information and searching for trends and patterns;
- making and testing hypotheses;
- refining and redefining ideas;
- reflecting on issues.

All these attributes, of course, contribute to good practice in learning mathematics for all children. ICT provides opportunities to enhance the role of those features in the classroom and offers a strategy for differentiating for the more able mathematicians who, as research tells us, are able to use these processes with greater competence.

Using structured programmes

Software packages which are structured in such a way as to enable pupils to practise and consolidate ideas should be used with greater care when working with very able pupils. But there are programs, such as 'Guess', 'Number line' and so on (SMILE resources) which have different levels of difficulty built into them so as to make differentiation easier to achieve.

Websites

An increasing number of mathematics websites which can be used to provide support for promising mathematicians are being set up. Some of them provide mathematical investigations, puzzles and challenges which can enrich pupils' mathematical experiences. Some useful websites are included in the list of resources in Chapter 7.

Summary

In this chapter I have considered the role of ICT in providing enriching mathematical experiences for able pupils. The main focus has been on the factors which make ICT particularly suitable when considering provision for able pupils. Examples of how calculators and computers can be used to enrich the mathematical learning of pupils were provided. These examples are based on what I have seen or experienced in classrooms. So these ideas should be both useful and manageable.

CHAPTER 7

Selecting resources for mathematically able pupils

Before discussing aspects of selecting and using resources for the mathematically able, we should perhaps examine what criteria should be used to select a suitable curriculum for such children. Many of the points raised in this connection have been touched on in previous chapters, but are worth revisiting here. This will not only remind us of what we should be aiming at, but will also help us to set up a mechanism for evaluating the resources. I feel that resources are a means to an end and, in the context of this book, the resources we select should enable us to satisfy the criteria for excellence in provision for mathematically promising pupils.

In judging what may be an appropriate curriculum for the able mathematician, the following pointers should be of help. These are principles of good practice for all children, but especially for able students.

- Curriculum materials should encourage the investigation of ideas, questioning and open-ended enquiry.
- Activities should encourage hypotheses, conjectures and generalisations.
- There should be a balance of facts, skills and problem-solving strategies.
- The curriculum should encourage children to go beyond what is being asked in a given task. This will enable them to make connections with previously learnt ideas and also appreciate the usefulness and relevance of mathematics in real life.
- Cut down on the amount of computational practice and make time for in-depth explorations.
- Ensure that children experience all areas of mathematics and do not concentrate only on number work.
- Make use of the power and versatility of ICT as much as possible.
- Encourage children to take part in competitions and enrichment clubs.

- Let the able children take a lead in designing books of puzzles, writing their own personalised textbook pages and diaries, conducting surveys and taking responsibility for displays of 'best work'.

Selection and choice of resources

It may seem strange that I have included the teacher here as a resource. My justification for doing so is that I believe that the teacher is the most valuable resource in a classroom. So I feel it is important that we start with a reflection on the attributes of an effective teacher of mathematically able pupils. Before reading on, try to make a list of what those attributes may be. What I have included in the following list has emerged from much discussion with teachers on in-service courses focusing on provision for mathematically able pupils. So, what are those characteristics?

The characteristics of teachers of the mathematically able

The following features are desirable in teachers who are committed to making the best provision for able pupils.

Flexibility
The word 'flexibility' is used here to focus on several aspects. Teachers need to be flexible in the identification of mathematically promising pupils. They need to use a range of methods of identification, using some of the methods described in Chapter 2, and need to adopt suitable styles of teaching and be prepared to let children work in groups or individually. Curriculum materials need to be selected to match the needs of the pupils and some curriculum compacting may be necessary to avoid the able pupil feeling frustrated and bored.

Interested in personal development
Teachers of the mathematically able will demonstrate an enthusiasm for the subject in order to inspire the pupils. They may be subject specialists or would have attended in-service courses to update their own knowledge and current thinking. They would be interested in recent developments and research relating to provision for able pupils.

Striving for excellence

Teachers will recognise that excellence in teaching and learning should be achieved at all times. In order to make excellence a habit, they will set high expectations, provide high quality resources and organise displays of pupils' work which are of a very high standard.

Textbooks will be used selectively and sensitively.

Encouraging reflection

Reflecting on mathematical ideas will support children in the construction of a robust framework of understanding. Teachers will encourage mathematical diaries, posters and interactive displays. Making presentations to others helps children to feel confident and develop self-esteem.

Learning in partnership

Teachers should be willing to admit that they may not know all the answers, but are happy to work with the children. They should not feel undermined by children who may be able to process information faster and ask smart questions. They should be willing to show that the students provide new insights into the subject. Having a sense of humour and being able to create a relaxed atmosphere will let the children feel at ease and want to excel in the subject.

Some ways of supporting the mathematically promising

If you have mathematically able pupils in your school, there are many ways in which you can provide for their needs. Here are some helpful suggestions.

In the classroom

Have puzzles and problem-solving materials available. Problems should include those with multi-level solutions. Investigations should be open-ended so that pupils can undertake explorations at their own level. It is not necessary for all children to be given the same starting point.

Attending 'pull-out' programmes

Your school may arrange a special club or enrichment time for the brightest pupils outside the mathematics lesson. These programmes should be targeted at the most able mathematicians and at those who have the potential to do very well in mathematics. Some encouragement may be needed for those able pupils who may be reluctant to join these programmes.

Specialist schools

It is worth remembering that some secondary schools have specialist expertise in mathematics. These schools will have a high profile for the subject and the staff are likely to have a high level of interest and considerable expertise in the subject. They may be interested in arranging programmes for bright young mathematicians in the local area.

Saturday or summer programmes

Your local authority may have master classes for able mathematicians. Increasingly, universities are also organising enrichment programmes for able mathematicians. Some charity organisations (listed later) also run Saturday or summer programmes. All LEAs now receive funding from the DfEE for summer school programmes for 'gifted and talented' pupils.

World-class tests

From 2001, pupils will be able to take world-class tests for 9- and 13-year-olds in mathematics and in problem solving. These tests are set in such a way that they will test the brightest using international standards. The Qualifications and Curriculum Authority (QCA) will provide information.

Mathematics clubs on the Internet

Projects such as NRICH, which is listed under the resources, provide excellent opportunities for able pupils to undertake mathematical investigations and share ideas. This website has been particularly commended by many teachers and by pupils who use it regularly.

Set up your own mathematics club

You could set up a mathematics club in your school and let the pupils take the responsibility for running it. Adults will need to be present, but organising a mathematics club is a highly motivating and challenging activity. It is particularly challenging if able pupils from other schools in the same local area join the club. Enlist the help of parents and other local experts who have mathematical expertise. The local library may also want to be involved.

Is your school providing support for your able mathematicians?

Most schools have some arrangements for supporting their most able mathematicians. These may not always be known to all the staff. Having a policy statement which explains what the school's commitment is and what is available would be very helpful. Most mathematics policies include a statement on provision for pupils with learning difficulties; it is only fair that a policy is in place for providing for the able mathematicians. Before designing a policy it would be useful to undertake an audit of existing provision.

Here is a possible checklist you could use for an audit.

- Does the school have any guidelines covering the needs of able pupils?
- What methods are used for identifying pupils who have high mathematical ability?
- Are there any special arrangements for organising teaching for the cohort of able pupils?
- What steps are taken to consider underachievement of pupils with high potential?
- Does the school enlist the partnership of parents in identifying talent and in making effective provision?
- Does the school have support from any external agencies?
- What is done to celebrate the able pupils' possible exceptional achievement?
- Has any staff development taken place in recent years?
- Is a list of resources made available to all staff?

Based on the audit, action needs to be taken. A policy also needs to be designed and disseminated. The policy needs to be concise, succinct and clear. Possible headings for the policy may include: rationale; aims; identification methods; provision (include special features of the curriculum and organisational arrangements); teaching styles; special programmes; external support; and resources.

Resources

The following are useful resources for mathematically able pupils.

- *Bright Challenge*, Casey, R. & Koshy, V. (Stanley Thornes, 1995) consists of activities for able pupils in mathematics, English and science. It can also be obtained from the Brunel Able Children's Education Centre, School of Education, Brunel University, 300 St. Margaret's Road, Twickenham TW1 1PT.

- *Extending Mathematical Ability Through Whole Class Teaching*, Fielker, D. (Hodder & Stoughton, 1997) is a very useful book which provides activities and examples of good practice.

- Also consider *Mental Mathematics* – ages 5–7, 7–9 and 9–11, Koshy, V. (Collins Publications, 1998), *Effective Teaching of Numeracy*, Koshy, V. (Hodder and Stoughton, 1999) and *Mathematics for Primary Teachers*, Koshy, V., Ernest, P. & Casey, R. (Routledge, 2000).

Non-scheme based resources

The following resources are available from the Association of Teachers of Mathematics (ATM) Publications, 7 Shaftesbury Street, Derby DE23 8YB (Tel: 01332 346 599):

- *Exploring Mathematics with Younger Children*
- *Points of Departure Books 1 to 4*
- *Teaching, Learning and Mathematics with IT*
- *Teaching, Learning and Mathematics: Challenging Beliefs*

Excellent materials are available from Tarquin Publications, Stradbroke, Diss, Norfolk, IP21 5JP (Tel: 01379 384 218).

For Cambridge Primary Mathematics (Key Stages 1 and 2), extension activities and puzzle pack, talking points and other useful materials, write to Cambridge University Press, The Edinburgh Building, Shaftesbury Road, Cambridge CB2 2RU.

Mathematics Pentathlon, PO Box 20590, Indianapolis, Indiana 46220, (317) 926-MATH

Journals

Cira – the Mathematical Magazine (highly recommended), is published by Juliet and Charles Snape Ltd., London NW6 1TH (Tel: 0207 433 1231).

Mathematics in School is published by the Mathematical Association, 259 London Road, Leicester LE2 3BE.

Mathematics Teaching is available from the ATM, 7 Shaftesbury Street, Derby DE23 8YB (Tel: 01332 346 599).

Computer resources

Mindstorms by Seymour Papert is a very useful book. Papert argues powerfully how problem solving using Logo helps children's mathematics learning. Logo is particularly useful for working with able pupils of Key Stages 1 to 3.

The British Educational Communications and Technology Agency (BECTA; formerly NCET), www.ncet.org.uk, provides a very useful list of support systems for using ICT; BECTA, Millburn Hill Road, Science Park, Coventry CV4 7JJ.

SMILE provides a very good range of software to support numeracy. The programs are differentiated at levels of challenge and provide feedback; SMILE, Isaac Newton Centre, 108a Lancaster Road, London. W11 1QS.

Websites

- NRICH (nrich.maths.org.uk) is an online mathematics club which should be of particular help for extensions activities.
- *Mathematics Ideas* (www.teachingideas.co.uk/maths/contents.htm)
- *Allmath.com* (www/allmath.com)
- *Interactive Mathematics*: miscellany and puzzles (www.cut-the-knot.com/content.html)
- *Maths Net* (www/anglia.co.uk/education/mathsnet)
- *Math Sphere* (www.mathsphere/co.uk)
- *Centre for Innovation in Mathematics Teaching* (www.ex.ac.uk/dmt)
- *Don Cohen, 'The MathMan'* (www.shout.net/~mathman) includes such resources as 'Calculus for Seven-Year-Olds' for precocious and motivated young mathematicians.
- *National Council of Teachers of Mathematics* (NCTM) (www/nctm.org)
- *Learning Websites for Middle School Students* (www-personal.umd.umich.edu/~jobrown/math.html)
- *AIMS* puzzle page (www.aimsedu.org/Puzzle/PuzzleList.html)
- Another site with challenging problems and information is www.cut-the-knot.com/
- *MATHCOUNTS*, National Society of Professional Engineers Information Centre (www.mathcounts.org/)
- *Mathematically Precocious Youth Program*, Johns Hopkins University, Baltimore, USA (www/jhu.edu/~gifted)

Support organisations

- The Brunel Able Children's Education Centre (BACE), School of Education, Brunel University, 300 St. Margaret's Road, Twickenham TW1 1PT; writes curriculum materials, researches into aspects of gifted education and provides training programmes on the 'gifted' for teachers of all Key Stages. Short courses and conferences are offered for coordinators and others.

- Children of High Intelligence, PO Box 4222, London SE22 8XG; runs Saturday classes and provides support for children and parents.

- Gift, The International Study Centre, Dartford Grammar School, West Hill, Dartford, Kent DA1 2HW; provides enrichment programmes for pupils, teacher support courses and materials.

- National Association for Able Children in Education (NACE), Westminster College, Oxford OX2 9AT; provides teacher support and a range of materials. NACE also provides short courses and conferences.

- National Association for Gifted Children (NAGC), Elder House, Milton Keynes MK9 1LR; provides support for parents. NAGC offers Saturday clubs and produces materials for teachers.

Summary

In this chapter I discussed aspects of selecting resources for the mathematically able. The attributes of effective teachers of the mathematically able were highlighted. A brief checklist was provided to facilitate an audit of provision in a school along with some useful headings for compiling a policy statement. The rest of the chapter was devoted to listing useful resources: books, packs of ideas, ICT based support and some useful websites.

References

Alexander, R. *et al.* (1992) *Curriculum Organisation and Classroom Practice in Primary Schools.* London: Department of Education and Science.

Association of Teachers in Mathematics (ATM) (1991) *Exploring Mathematics with Younger Children.* Derby: ATM.

Bell, A. W. *et al.* (1983) *A Review of Research in Mathematics Education: Part A, Teaching and Learning.* Windsor: NEFR-Nelson.

Bloom, B. S. (1956) *Taxonomy of Educational Objectives, Volume 1.* Harlow: Longman.

Burton, L. (1986) *Thinking Things Through.* Oxford: Blackwell.

Casey, R. (1999) 'A key-concepts model for mathematics learning', *Mathematics in Schools*, May 1999.

Casey, R. (2000) 'Whole numbers', in Koshy, V., Ernest, P. and Casey, R. (eds) *Mathematics for Primary Teachers.* London: Routledge.

Casey, R. and Koshy, V. (1995) *Bright Challenge.* Cheltenham: Stanley Thornes.

Centre for Talented Youth (1994) *Philosophy and Program Policy.* Baltimore, MD: Johns Hopkins University.

Cockcroft, W. H. (1982) *Mathematics Counts: Report of the Committee of Inquiry into the Teaching of Mathematics in Schools.* London: HMSO.

Department for Education and Employment (DfEE) (1995) *Mathematics in the National Curriculum.* London: HMSO.

Department for Education and Employment (DfEE) (1997) *Excellence in Schools.* London: DfEE.

Department for Education and Employment (DfEE) (1998) *The Implementation of the National Numeracy Strategy: The Final Report of the Numeracy Task Force.* London: DfEE.

Department for Education and Employment (DfEE) (1999a) *Excellence in Cities.* London: DfEE.

Department for Education and Employment (DfEE) (1999b) *The Framework for Teaching Mathematics from Reception to Year 6.* London: DfEE.

Department for Education and Employment (DfEE) (1999c) 'Morris welcomes Commons support for Gifted Plans', *DfEE News*, April.

Department for Education and Employment (DfEE) (2000a) *The Framework for Teaching Mathematics for Year 7.* London: DfEE.

Department for Education and Employment (DfEE) (2000b) *National Literacy and Numeracy Strategies: Guidance on Teaching Able Children.* London: DfEE.

Department of Education and Science (DES) (1988) *Mathematics for Ages 5 to 16. Proposals of the Secretary of State for Education and Science and the Secretary of State for Wales.* London: HMSO.

Department of Education and Science (DES) (1991) *Mathematics in the National Curriculum.* London: HMSO.

Ernest, P. (2000) 'Teaching and learning mathematics', in Koshy, V. *et al.* (eds) *Mathematics for Primary Teachers.* London: Routledge.

Fielker, D. (1997) *Extending Mathematics Ability through Whole Class Teaching.* London: Hodder and Stoughton.

Flavell, J. H.(1985) *Cognitive Development.* Englewood Cliffs, NJ: Prentice-Hall.

Freeman, J. (1998) *Educating the Very Able. Current International Research.* London: OFSTED.

Gardner, H. (1983) *Frames of Mind.* New York: Basic Books.

Gardner, H. (1993) *Multiple Intelligences.* New York: Basic Books.

George, D. (1990) *The Challenge of the Able Child.* London: David Fulton Publishers.

Gross, M. (1999) 'From "the saddest sound" to the D Major chord'. Keynote address presented at the 3rd Biennial Australasian International Conference on the Education of Gifted Students, August 1999, University of South Wales, Sydney.

Her Majesty's Inspectorate of Schools (HMI) (1978) *Primary Education in England: A Survey by HM Inspectors of Schools.* London: HMSO.

Her Majesty's Inspectorate of Schools (HMI) (1979) *Aspects of Secondary Education in England.* London: HMSO.

Her Majesty's Inspectorate of Schools (HMI) (1985) *Mathematics 5–16: Curriculum Matters 3.* London: HMSO.

Her Majesty's Inspectorate of Schools (HMI) (1992) *Education Observed: The Education of Very Able Pupils in Maintained Schools.* London: HMSO.

Hughes, M. *et al.* (2000) *Numeracy and Beyond.* Buckingham: Open University Press.

Koshy, V. (1998) *Mental Maths Teachers' Book 9–11.* London: Collins.

Koshy, V. (1999) *Effective Teaching of Numeracy for the National Mathematics Framework.* London: Hodder and Stoughton.

Koshy, V. and Casey, R. (1997) *Effective Provision for Able and Exceptionally Able Children.* London: Hodder and Stoughton.

Koshy, V. and Casey, R. (2000) *Special Abilities Scales.* London: Hodder and Stoughton.

Koshy, V. and Dodds, P. (1995) *Making IT Work for You.* Cheltenham: Stanley Thornes.

Koshy, V. *et al.* (eds) (2000) *Mathematics for Primary Teachers.* London: Routledge.

Kulik, J. A. and Kulik, C. A. (1992) 'Meta-analysis of evaluation findings on groupings programs', *Gifted Child Quarterly* 36(2), 73–6.

Krutetskii, V. A (1976) *The Psychology of Mathematical Abilities in School Children.* Chicago, IL: University of Chicago Press.

London Mathematical Society (1995) *Tackling the Mathematics Problem.* Report by the London Mathematical Society, the Institute of Mathematics and its Applications and the Royal Statistical Society. www.lms.ac.uk/policy/tackling/report.html (accessed 5 October 2000).

Mitchell, C. and Koshy, V. (1995) *Effective Teacher Assessment: Looking at Children's Learning.* London: Hodder and Stoughton.

National Council of Teachers of Mathematics (NCTM) (1980) *An Agenda for Action. Recommendations for School Mathematics of the 1980s.* Reston, VA: NCTM.

National Curriculum Council (NCC) (1989) *Non-statutory Guidance for Mathematics.* York: NCC.

National Curriculum Council (NCC) (1992) *Using and Applying Mathematics Book B – Inset Handbook*. York: NCC.

Office for Standards in Education (OFSTED) (1994a) *Exceptionally Able Children. Report of Conferences*. London: DfEE.

Office for Standards in Education (OFSTED) (1994b) *Science and Mathematics in Schools. A Review*. London: HMSO.

Office for Standards in Education (OFSTED) (1995) *Mathematics. A Review of Inspection Findings. 1993/94*. London: HMSO.

Office for Standards in Education (OFSTED) (1998) *Setting in Primary Schools*. London: OFSTED.

Ogilvie, E. (1973) *Gifted Children in Primary Schools*. London: Macmillan.

Papert, S. (1980) *Mindstorms: Children, Computers and Powerful Ideas*. New York: Basic Books.

Polya, G. (1945) *How to Solve It* . Princeton, NJ: Princeton University Press.

Renzulli, J. (1986) 'The three-ring conception of giftedness: a developmental model for creative productivity', in Sternberg, R. J. and Davidson, J. E. (eds) *Conceptions of Giftedness*. Cambridge: Cambridge University Press.

Renzulli, J. and Reis. S. (1994) *Schools for Talent Development: A Practical Plan for School Improvement*. Mansfield Center, CT: Creative Learning Press.

Resnick, L. B. (1987) *Education and Learning to Think*. Washington, DC: National Academy Press.

Sheffield, L. (1994) *The Development of Gifted and Talented Mathematics Students and the National Council of Teachers of Mathematics Standards*. Connecticut: The National Research Centre on the Gifted and Talented.

Sheffield, L. (1999) 'The development of mathematically promising students in the United States', *Mathematics in School* **28**(3).

Sheffied. L. *et al*. (1995) 'Report of the task force on the mathematically promising', *NCTM Bulletin*, 32, December.

Skemp, R. (1976) 'Relational understanding and instrumental understanding', *Mathematics Teaching*, December.

Straker, A.(1982) *Mathematics for Gifted Pupils*. Harlow: Longman.

Sutherland, R. (1995) 'Symbolising through spreadsheets', *Teaching and Learning Mathematics with IT*. Derby: ATM.

Terman, L. M. and Oden, M. H. (1947) *Genetic Studies of Genius – IV: The Gifted Child Grows Up*. Palo Alto, CA: Stanford University Press.

Van Tassel-Baska, J. (1992) *Planning Effective Curriculum for Gifted Learners*. Denver, CO: Love Publishing Company.

Vygotsky, L. (1978) *Mind in Society*. Cambridge, MA and London: Harvard University Press.

Index